MANUSCRIPTS OF THE DIBNER COLLECTION

SMITHSONIAN INSTITUTION LIBRARIES
Research Guide No. 5

MANUSCRIPTS OF THE DIBNER COLLECTION

IN THE

Dibner Library of the History of Science and Technology

OF THE

SMITHSONIAN INSTITUTION LIBRARIES

◄§ Smithsonian Institution Libraries §►
Washington, D.C.
1985

Publication consultant and sole distributor
WATSON PUBLISHING INTERNATIONAL
Canton, Massachusetts 02021

Library of Congress Cataloging in Publication Data

Smithsonian Institution. Libraries.
  Manuscripts of the Dibner collection.

  ([Smithsonian Institution Libraries research guide ; 5])
  Includes index.
  1. Science—History—Sources—Bibliography—Catalogs.
2. Science—History—Manuscripts—Catalogs.
3. Technology—History—Sources—Bibliography—Catalogs.
4. Technology—History—Manuscripts—Catalogs.
5. Dibner Library—Catalogs. I. Dibner, Bern.
II. Title. III. Series.
Z7405.H6S54 1985   [Q125]   016.5   ·85-11576
ISBN 0-88135-025-7
ISSN 0732-7447

CONTENTS

LIST OF ILLUSTRATIONS

To
Bern Dibner
who assembled this remarkable collection

Bern Dibner photographed with his wife Barbara,
November 1983.

Foreword

This collection of manuscripts from the Dibner Library of the History of Science and Technology was assembled by Dr. Bern Dibner over a period of nearly five decades and presented to the Smithsonian Institution in 1976. The 1,614 groups of manuscripts which reflect nearly 900 years of developments in western science and technology are currently housed in the principal special-collections facility of the Smithsonian Institution Libraries, located in the National Museum of American History. Many of the most distinguished names in the history of science and technology from the High Middle Ages to the present are represented. This deposit, therefore, reflects a rich and varied picture of our scientific and technological heritage. This volume is the fifth in the Smithsonian Institution Libraries Research Guide Series and we publish it as an invitation to students of the many disciplines and periods represented to come and study the collection, confident that further research will inspire vigorous scholarly studies and thereby promote the "increase and diffusion of knowledge" as mandated by Smithson.

Manuscript items range from single parchments and bound manuscript books of the medieval period to lecture notes and typescripts of this century. To illustrate the wide range of works here represented, two late medieval scientific works are worthy of note—a late-thirteenth-century copy of the *De proprietatibus rerum* of Bartholomaeus Anglicus and an early-fourteenth-century edition of Sacro Bosco's *De sphaera*. Another fourteenth-century holding is an Averroist commentary on Aristotle's *Physics*, one of several scholastic manuals and commentaries. Two books yielded fragments of medieval manuscripts used as end leaves. One, dating from the twelfth century, is probably from a collection of sermons while the second is a fourteenth-century liturgical manuscript containing the *Canticum Moysi*. A fifteenth-century copy of Boethius' *De arithmetica* written in a very fine humanistic bookhand is among the several notable items illustrating the development of scripts.

The Renaissance period is represented by such scholars as Philipp Melanchthon, Peter Ramus and Johann Stabius. Robert Boyle, Tycho Brahe and René Descartes are among the prominent figures associated with the dawn of modern science whose manuscripts are housed here, along with Johannes Kepler whose papers include his calculations and fragments of his *Eclogae Chronicae*. Several manuscripts of the seventeenth and eighteenth centuries trace the development of the barometer. Scientific correspondence of the early modern period includes a letter from Galileo to the French astronomer Pieresc and letters from Gilbert Clerke to Isaac Newton, one with a reply from Newton on problems in his *Principia*. Other papers of Newton include several studies on alchemy and chemistry.

Dr. Dibner's particular interest in the history of electricity is reflected in many of the eighteenth- and nineteenth-century manuscripts. Benjamin Franklin, Alessandro Volta, and Michael Faraday are among the pioneers in this field whose works are represented. Researchers of the donation will also find ninety-nine of the letters exchanged between Abbé Nollet and Etienne Dutour, an unpublished manuscript, *Essai d'Histoire de la Physique* (1814), by Jean André Deluc, and drawings of flasks from the laboratory of Louis Pasteur. The prolific correspondence of Benjamin Silliman and Silvanus Thompson is housed here as well. An example of the interplay between science and politics is found in a proposal made to Napoleon I (1811) for the translation into French of the proceedings of Galileo's trial, the records of which Napoleon had seized from the Vatican Archives.

Among the more recent items, there are over 100 letters with the papers of Ernst Mach, a group of Max Planck's letters, a copy of the Nobel lecture delivered by Marconi, a large number of Albert Einstein's papers, a manuscript on ferromagnetism by Werner Heisenberg, and two scientific articles in typescript by Jonas Salk. Manuscripts pertaining to the history of aeronautics range from a letter of Orville Wright to two articles on helicopters by Igor Sikorsky and a brief description by John Glenn of preparations for orbital flight.

Several of the manuscripts have been published as facsimiles. The letter written by Galileo to Peiresc in 1635 on the magnetic clock (Item 58) was reproduced and discussed in its historical context by Stillman Drake in B. Dibner and S. Drake, *A Letter from Galileo* (Norwalk, Conn.: Burndy Library, 1967), pp. 45–56. An illustration of the drawing made by Martinus van Marum suggesting changes for the improvement of Teyler's static electricity machine (Item 396) appeared in Dibner's *Early Electrical Machines* (Norwalk, Conn.: Burndy Library, 1957). Dr. Dibner has also published a facsimile of Michael Faraday's 1831 letter to Richard Phillips (Item 686) in his *Faraday discloses Electro-magnetic induction: His epochal letter sent from Brighton to Richard Phillips, F.R.S., is here reproduced* (New York: Burndy Library, 1949) as well as a facsimile page from Chapter VII of the *Origin of Species* (Item 747)

in Darwin's hand (B. Dibner, *Darwin of the Beagle* (Norwalk, Conn.: Burndy Library, 1960). A comment on the attribution to Ambroise Bachot of the Ramelli drawings was published by Martha T. Gnudi in *Technology and Culture* XV (1974):614–25.

This published list provides basic bibliographic information, generally following guidelines set out in the second edition of the *Anglo-American Cataloguing Rules* (1978). Entries are arranged in rough chronological order, beginning with the earliest manuscripts. As a single entry often represents a number of items within a fairly broad range of dates, the chronological order is not strict. Maps, tables, and charts have not been catalogued separately, but are grouped under their respective entries. Illustrations included in this book are cited in bracketed roman numerals. The library call number is listed as the final citation. Captions to the illustrations cite the negative number of the Smithsonian Institution (SI) Office of Photographic Services. The guide appends an index of subjects, of titles and authors, and of personal names of individuals listed in both the main entries and in the notes. All index listings refer to the item number in this published catalogue. The manuscript citations have been entered into OCLC (Online Computer Library Center), a nationally and internationally accessible data base. The *National Union Catalogue Pre-1956 Imprints* (Washington, D.C., 1968–81) has been used as the authority for the form of names and titles. Biographical details have been confirmed, where possible, in the *Dictionary of Scientific Biography* (New York, 1975), J. C. Poggendorf's *Biographisch-Literarisches Handwörterbuch der Exakten Naturwissenschaften* (Leipzig, Berlin, 1863-   ), and the *Dictionary of National Biography* (Oxford, 1917-   ).

Members of the SIL staff who worked on the cataloguing of these manuscripts benefited from information provided by D. G. C. Allan of the Royal Society for the Encouragement of Arts, Manufactures and Commerce, London; F. Dolbeau of the Institut de Recherche et d'Histoire des Textes, C. N. R. S., Paris; Julia Elton of B. Weinreb, Ltd., London; Robert A. Hatch of the University of Florida, Gainesville; T. Koehler, S.M., of the Marian Library, University of Dayton; Letje Multhauf of Washington, D. C.; L. J. Reilly of Woodlands, the Local History Library of the London Borough of Greenwich, England; N. H. Robinson of the Royal Society, London; Gerhard Stamm of the Badische Landes-Bibliothek, Karlsruhe, Germany; D. G. Vaisey, Keeper of Western Manuscripts, Bodleian Library; and William Wallace of The Catholic University of America. We wish to thank all these individuals for their assistance. Any remaining errors in fact or judgment are, naturally, borne by the Libraries. We welcome emendations to the information printed from all future users of the collection.

The publication of this catalogue list would not have been possible without the critical financial support the Libraries received from the following sources: The Dibner Fund, Inc., of Norwalk, Connecticut, generously provided funds for cataloguing and for other costs, The Atherton Seidell Endowment Fund of the Smithsonian Institution awarded the Libraries a grant to support production, and the Smithsonian Women's Committee supplied monies to supplement costs for additional personnel. We thank those organizations for their important contributions to this project.

We would like to express our gratitude to Mrs. Lorie Aceto, Deputy Director, Smithsonian Institution Office of Printing and Photographic Services, who photographed the manuscripts appearing in this volume. The charming photograph of Dr. and Mrs. Bern Dibner was taken by their grandson, Daniel Dibner.

It is always a pleasure to note our appreciation for the dedicated volunteers who help the Smithsonian Libraries in so many crucial tasks. In this undertaking we are indebted to Filomena Darroch, Margaret Dong, Anne Eales, Linda Livingston, Leslie Overstreet, Hugh Pettis, George Steyskal, and Betty Youssef.

Numerous members of the Smithsonian Institution Libraries staff cooperated in this venture from the moment the manuscripts first arrived in 1976 to the final proofreading in the last pre-production stages. The initial sorting was the responsibility of Ellen B. Wells, Chief Librarian of the Special Collections Branch Library, assisted by Mary Augusta Rosenfeld; both provided expert advice throughout the project. Cataloguers Dianne Chilmonczyk, Sigrid P. Milner, and Catherine Brown Tkacz accomplished their tasks under the capable direction of Mary Jane Linn, Chief of Original Indexing Services, and Vija Karklins, Deputy Director. Laura A. Bedard aided in the cataloguing efforts. Johannes H. Hyltoft, Chief Conservator, skillfully ensured the physical integrity of several manuscripts and Edward Johnson of the Special Collections Branch Library assisted with the delicate photographic operations. The text of this work was entered on computer disks by Paulette E. Gaskins, Publications Program Assistant, and Nancy L. Matthews, the Libraries' Publications Officer, was responsible for publication arrangements. Finally, this project owes a special debt to Michael W. Tkacz who coordinated the final stages of research and arrangement and took responsibility for editing the entire manuscript. We take pleasure in dedicating this volume to Bern Dibner who created this remarkable collection as part of the Dibner Library of the History of Science and Technology which is now available for further study by scholars and the public.

Robert Maloy, Director
Smithsonian Institution Libraries, 1985

ABBREVIATIONS

A.	autograph
D.	document
L.	letter
S.	signed, signature
T.	typed
A.L.	autograph letter
A.S.	autograph signature
D.S.	document signed
L.S.	letter signed
A.D.S.	autograph document signed
A.L.S.	autograph letter signed
T.L.S.	typed letter signed
ms.	manuscript
mss.	manuscripts
cent.	century
cm.	centimeter(s)
col.	color
ill.	illustration(s)
p.	page(s)
port.	portrait

1. *Sermons.* [late 12th century]. 2 sheets (4 leaves)

Italian parchment ms. fragments containing sermons. Both sheets are in the same hand and without decoration except a rubricated E beginning the text "Ecce odor filii mei sicut odor agri pleni cui benedixit dominus" (Gen. 27.27b). Script appears to be Italian and is probably of the late 12th century; in Latin. Found as end leaves in Theophilus de Ferrariis, O.P., *Propositiones ex omnibus Aristotelis libris* (Venice, 1493); cf. *Gesamtkatalog der Wiegendrucke,* n. 9826. The Ferrariis incunabulum contains the handwritten ownership note: "Liber hic pertinet conventum fratrum minorum de observantia. Ad sanctum Donatum prope Urbinum civitatem." [I]   MSS 1610 A

2. **Bartholomaeus Anglicus, fl. 1230-1250.** *[De proprietatibus rerum].* [last quarter of 13th century]. 38, 229 leaves (i.e. 76, 458 p.); 25 cm.

On vellum with rubricated initials; page one of text has two illuminated initial letters and marginal decorations with figures; in Latin. Occasional marginalia in a later hand. Stamp on leaf 47 recto: Biblioth. Privata, P. Praep. Gen. S.J. Index on first 35 leaves (leaves 1-2 lacking). On spine: Aegidius De Proprietatibus rerum, Codex.

MSS 241 B

3. **Averroës, 1126-1198.** *Com[m]e[n]tum av[err]oys sup[er] lib[r]o Phy[sic]a ar[ystoteles].* [14th century]. [324] p.; 42 cm.

Vellum ms. with rubricated initials; p. 1 has an illuminated initial letter; in Latin. Bound in folio sheet of music [1075-1125?].   MSS 288 B

4. *Canticum moysi.* [14th century?]. 1 item (1 leaf)

Manuscript leaf used as cover, with rubricated title and initials; in Latin. Found with: Alberto magno delle virtu dele herbe. [16th century?].   MSS 5 B

5. **Boethius, Anicius Manlius Severinus, 480-524.** *[De institutione arithmetica].* [15th century]. 76 leaves [i.e. 146 p.]: col. ill.; 31 cm.

Text written in a fine early humanistic bookhand with rubricated initials. The text of the treatise is complete and is accompanied by a number of diagrams and tables in color; in Latin.   MSS 286 B

6. **Sacro Bosco, Johannes de, fl. 1230.** *Tractatus de Sphaera Johannis de Sacrobosco Item Liber quidam sortilogus; quibus accedit De arithme[tica] et geometria Tractatus authore anonymo, 1637.* [1st half of the 15th century?]. [92] p.: ill. (some col.); 17 cm.

Rubricated ms., with several tables and diagrams; bound and title page added in 17th century; in Latin. Spine title: Sacrobosco Tractatus de sphera, England ca. 1430.   MSS 1272 B

7. **Aristotle, 384-322 B.C.** *Auctoritates extracte ex Libris phi[losophorum]; et p[rim]o ex libro p[rim]o metaphi[si]c[o]r[um].* [ca. 1450]. [92] p.; 16 cm.

Extracts from Aristotle and other philosophers dealing to a great extent with medicine, science, and metaphysics with rubricated initials and subtitles; in Latin.   MSS 273 B

8. **Cecco d'Ascoli, 1269-1327.** *[L'Acerba].* [mid-15th century]. 78 leaves [i.e. 156 p.]: col. ill.; 26 cm.

Vellum ms. with rubricated initials, some lines in red ink; p. 1 of text has illuminated initial letter and decorations; in Italian. On binding: Franco. Donato, MDLXII. On box: Cecco d'Ascoli, 1461.   MSS 279 B

9. Regiomontanus, Joannes, 1436-1476. *Paper.* [between 1456 and 1476]. 1 item (1 leaf)

A. ms. S.; in German.   MSS 1046 A

10. *Alberto magno delle virtu dele herbe delli animali pietre pretiose et di molto maraviliose cose del mondo dinove con diligeenza [sic] ristampato e corretto.* [16th century?]. [90] p.: ill.

Receipt book, in an unknown Italian hand, supposedly based on the *Liber aggregationis* ascribed to Albert, the first printed translation being Venice, 1502 (cf. *National Library of Medicine 16th Century Catalogue*); in Italian. Addendum: ms. recipe laid in p. [28]. With: Canticum moysi [14th century?], a ms. leaf used as cover.

MSS 5 B

11. *Drawings of machines.* [16th century]. 6 items: ill.

Pen and wash drawings on vellum, probably by Bachot, highlighted with touches of white, mounted on 18th century heavy paper. Possibly drawn for Ramelli's *Le diverse figure et artificiose machine* (1588), figures 131, 158, 169, 176, 181, 182, and 189. Without key lettering as published. In pencil at bottom of [181], on early mounting: 12. Da Filipo Bruneleschi. [II]   MSS 1604 A

12. Kratzer, Niklas, 1487-1550. *Letter.* [between 1507 and 1550]. 1 item (1 leaf)

A.L.S. (Oct. [5?], London); in German.   MSS 797 A

13. *Libro di aritmetica.* [first half of 16th century]. 202, 27 p.; 14 cm.

Treatise on arithmetic; in Italian. Spine title. Also on spine: M.S. cartageo del S. XVI, F.S.   MSS 219 B

14. Stabius, Johann, 1450-1522. *Paper.* 1508. 1 item (1 leaf)

A. note S. (1508 Feb. 15); in German.   MSS 1402 A

15. Grynaeus, Simon, 1493-1541. *Letter.* [between 1513 and 1541]. 1 item (1 p.)

A.L.S. (June 24, Basel) concerning the king of France; in Latin.   MSS 629 A

16. Tannstetter von Thannau, Georg, 1482-1535. *Letter.* 1513. 1 item (1 leaf)

A.L.S. (1513 April [16?]) to Stabius; in Latin.
MSS 1444 A

17. Virdung, Johann, ca. 1465-ca. 1535. *Letter.* 1525. 1 item (1 leaf)

A.L.S. (XXV, i.e., 1525, nach Leonhard), to Michael von Sensheim; in German. With detached seal.
MSS 1510 A

18. Melanchthon, Philipp, 1497-1560. *Letter.* 1538. 1 item (2 p.)

A.L.S. (1538 Mar. 3) to Theodore recommending a young man who would deliver the letter; in Latin.
MSS 987 A

19. *Systema physicum.* 1548-1631. [46] p.: ill.; 16 cm.

Rubricated ms., dated in colophon (die decimo tertio mensis decembris anno 1548, i.e., 1548 Dec. 13). Text and commentary in margins in one hand, additional marginal notes and title page in a later hand. Title page dated 1628 by same hand that wrote a note (1631 Mar. 10) inside back cover; in Latin.   MSS 1290 B

20. *Miscellanea alchymica et astrologica.* [2nd half of 16th century]. 104 leaves: ill. (some col.); 32 cm.

Several different texts, mainly alchemical and astrological but also concerning medicine, the properties of stones, the church calendar, etc. Rubricated, profusely illustrated. Also includes several shorter texts and notes on leaves 72-84, with dates up to 1597. Most articles and notes are in Latin, some in Italian. Dealer's description laid in. [III]   MSS 867 B

21. Leovitius, Cyprianus, d. 1574. *Letter.* 1551. 1 item (2 p.)

A.L.S. (1551 [Jan.?] 19); in German.   MSS 883 A

22. Peucer, Kaspar, 1525-1602. *Letters.* 1552-1593. 2 items.

A.L.S. (1552 Feb. 18) to Melanchthon on the dangers of the times and perseverance in teaching, and A.L.S. (1593 April 28) to a cleric named Hieronymus; in Latin.
MSS 1121 A

I. Parchment fragment of the 12th century containing comments on a biblical text.

Entry No. 1, SI Neg. 84-7185/6.

23. Luther, Paul, 1533-1593. *Document.* 1563-1848. 1 item (1 leaf)

A.D.S. (1563 May 11) containing Greek and Latin epigrams, with a holograph annotation (1848 Mar. 4) signed D. Spiehr(?); in German. MSS 932 A

24. Camerarius, Joachim, 1534-1598. *Papers.* [ca. 1565-ca. 1605]. 2 items.

A. note S. (ca. 1565) and ms. list of epitaphs composed by Camerarius (ca. 1605); in Latin. MSS 293 A

25. Ramus, Petrus, 1515-1572. *Papers.* 1569-1571. 2 items.

A.L.S. (1571 [Nov.?] 25, Paris) to Lavater about a discussion concerning Aristotle in which Bullinger and Stephan Diner are involved, and an autograph signature (1569) on sheet with note in another hand: ". . . Ramus à Wilhelmo Adolpho [Simbonio?]. . ."; in Latin. MSS 1188 A

26. Theodoricus, Sebastianus, fl. 1553. *Letter.* 1572. 1 item (1 leaf)

A.L.S. (1572, perhaps from Wittenberg) to Menius mentioning Eberhard Wolff and making a request; in Latin. Address cut off. MSS 1455 A

27. Baldi, Bernardino, 1553-1617. *Discorso di Leontio artefice sopra la sfera di Arato et fabrica di quella.* [between 1573 and 1617]. [31] p.; 21 cm.

Bound in 13th century vellum leaf with Latin text. Boncompagni ms. 339; cf. Narducci, E., *Catalogo di manoscritti,* 2nd ed., 1892. MSS 235 B

28. Zuccolo, Vitale, 1556-1630. *Dialogo delle cose meteo[ro]logiche di D. Vitale Zuccolo [?---ini] Monaco Camaldolense.* [1576-1690?]. [264] p.: ill.; 27 cm.

Bound ms. containing a text published in 1690. The characters in the dialogue are Battista Peroli, Stefano Viari, and Camillo Abbioso; in Italian with extensive Latin quotations. Spine title: Meteoro[logico] dialogo. MSS 1311 B

29. Erastus, Thomas, 1524-1583. *Letter.* 1578. 1 item (1 p.)

A.L.S. (1578 Mar. 25) to Lavater; in Latin. MSS 485 A

30. *An[n]otationes et quaestiones logicae.* [ca. 1600?]. [10], 176 p.

Notes on philosophical and scientific subjects, including logic, rhetoric, and poetics; in Latin. Imperfect copy: lacks p. 161-2 (replaced with p. 179-[180]), p. 173-4 and p. 177-8. MSS 67 B

31. Gregory XIII, Pope, 1502-1585. *Document.* 1585. 1 item (1 p.)

D.S. (1585 April 2) concerning inheritance from Paulo Odescalco, bishop of Tonne; in Italian. MSS 618 A

32. Praetorius, Johann, 1537-1616. *Letter.* 1587. 1 item (1 leaf)

A.L.S. (1587 June 6, [Altdorf?]) to Hieronymus Pangartner mentioning Erasmus Wagner and concerning the education of a young man and the value of the example of excellent men; in Latin. MSS 1169 A

33. *Kunstbuech der Puchsenmeisterey [sic].* 1589. [120] leaves: ill.; 30 cm.

Treatise on master-gunnery. Manuscript bound and trimmed in 17th century, clipping text from top or side of several leaves. Numerous ink drawings, mostly of quadrant, pasted in last 20 leaves. Final drawings folded; in German. Dealer's description, in German, laid in. [IV] MSS 835 B

34. Brahe, Tycho, 1546-1601. *Note.* 1590. 1 item (1 p.)

A. note S. (1590 July 20); in Latin. MSS 163 A

35. Kepler, Johannes, 1571-1630. *Papers.* [between 1591 and 1630]. 3 items.

D.S. in German, sheet of autograph calculations (4 p.), and holograph ms. (4 p.) of excerpt from *Eclogae Chronicae,* in Latin and Greek. Additional materials include facsimile, typescript copy and ms. copy of parts of the *Eclogae Chronicae.* Photocopies in second folder. MSS 782 A

36.　Ortelius, Abraham, 1527-1598. *Letter.*
1597. 1 item (1 leaf)

A.L.S. (1597 April 18, Antwerp) to Camerarius; in
Latin.　　　　　　　　　　　　　　MSS 1090 A

37.　Chytraeus, David, 1531-1600. *Letter.*
1599. 1 item (3 p.)

L.S. (1599 Feb. 26, Rostock) to Crusuis and Zellius on
Chytraeus's seventieth birthday, with holograph draft; in
Latin.　　　　　　　　　　　　　　MSS 346 A

38.　Constantius Philalis(?), 17th cent. *Philosophia universa in tres tomos divisa auctore Constantio Philalite, cosmopolitano.* 1620. [711] p.:
ill.; 26 cm.

Half-sheet of notes and calculations laid in. Book plate:
Ex libris Robert Billecard.　　　　　　MSS 275 B

39. *De scientia astrorum.* [17th century].
[444] p.: ill.; 28 cm.

Treatise on astronomy; in Latin. Spine title, with date
1650.　　　　　　　　　　　　　　MSS 259 B

40. *De spiritalibus.* [17th century?]. [99] p.: ill.
(some folded); 18 cm.

Some diagrams tipped in; many blank leaves; notes on
back end-paper; in Latin and Italian.　　MSS 255 B

41. *[L'Idrostatica].* [17th century]. [12] p.; 22
cm.

Holograph ms. on hydrostatics; in Italian.
　　　　　　　　　　　　　　　　MSS 230 B

42. *Machine et instrumenti de piu celebratissimi autori.* [17th century]. 40 leaves: all ill.; 34 cm.

Forty ink wash drawings of water wheels, pulleys, and
other machinery planned by Vitruvius, Besson, Capra,
Meyer, and seven others.　　　　　　MSS 864 B

43.　Profidus, Marcus Antonius. *[Discorso della philosofia naturale e morale].* [17th century?]. [99] p.; 19 cm.

Holograph ms. in Italian with Latin colophon.
　　　　　　　　　　　　　　　　MSS 1256 B

44.　Nautonier, Guillaume de, sieur de Castelfranc, fl. 1603. *Letter.* 1604. 1 item (2 p. in 4
pieces)

A.L.S. (1604 May 15, Paris) possibly to Scaliger; in
French.　　　　　　　　　　　　　MSS 1059 A

45.　Mayer, Daniel, 16th-17th cent. *Gründtlicher Bericht vom Gebraüch des gevierten geometrischen Instruments alle höhe weite lenge und teiffe abzumessen Dessgleichen vom Feldmessen wie man den inhalt eines jeden stück feldes rechnen und finden soll beschrieben durch Daniel Mayern.* 1607. 102 leaves: ill. (some col.);
10 x 15 cm.

Holograph on surveying. Illustrations include a colored
"Scala Altimetra," several diagrams, and a squareroot
table. Mayer's name and the date 1607 are on the first leaf.
On cover: D M V / 1607; in German.　　MSS 907 B

46.　Scaliger, Joseph Juste, 1540-1609. *Letter.* 1607. 1 item (2 p.)

A.L.S. (MDCVII, xvi kal. Februaris Juliani, i.e., 1607
Jan. 17) to Drusius concerning books and the correct
order of the name Sulpicius Severus; in Latin.
　　　　　　　　　　　　　　　　MSS 1328 A

47.　Cankler, Bernhard, 17th cent. *Epigram.*
1615. 1 item (1 p.)

Signed and dated holograph; in Latin.　MSS 296 A

48.　Descartes, René, 1596-1650. *Letter.* [between 1616 and 1650]. 1 item (1 p.)

A.L.S. written enclosing something in case Descartes
cannot present it in person; in French.　MSS 431 A

49.　Galilei, Vincenzo, 1606-1649. *Papers.*
1619-1649. 3 items.

A.L.S. (1649 Jan. 13) to Giovanni Nardi in Italian; a
holograph article "Cosmus" (1619 June 25, Florence), a
supposed copy of a Galileo ms. in Latin; and an A.D. of
accounts (undated) in Italian.　　　　MSS 561 A

50. Bernegger, Matthias, 1582-1640. *Papers.* 1621-1625. 2 items.

Two holograph quotations (one in Latin, one in Greek) with dedications and signatures in Latin.   MSS 94 A

51. Dudley, Robert, Sir, 1574-1649. *Letter.* 1622. 1 item (1 p.)

A.L.S. (1622 Sept. 25); in Italian.   MSS 457 A

52. Maria Christina, consort of Ferdinand I, de'Medici, grandduke of Tuscany, 1565-1636. *Letter.* 1622. 1 item (3 p.)

A.L.S. (1622 July 16, Florence) to Senatore Vettori; in Italian.   MSS 345 A

53. Pontanus, Johannes Isacius, 1571-1639. *Letters.* 1623-1637. 5 items.

Five A.L.S. (1623-1637, Harderwijk), 4 to Hoffer, the last to Hoffer's son; in Latin.   MSS 1159 A

54. Gassendi, Pierre, 1592-1655. *Letter.* [between 1625 and 1655]. 1 item (2 p.)

A.L.S. (undated) to Ismael Bulliardus (i.e. Ismael Boulliau); in Latin.   MSS 570 A

55. Maestlin, Michael, 1550-1631. *Autograph signature.* 1627. 1 item (1 leaf)

A.S. with Latin epigram (1627 May 8, Tübingen).   MSS 1017 A

56. Hofmann, Heinrich, 1576-1652. *Letter.* 1631. 1 item (2 p.)

A.L.S. (1631 May 30, Jena), with seal, contains Latin quotation and anagram on the name "Thomas Fincke"; in German. Additional material includes partial transcription.   MSS 711 A

57. Riccardi, Gabriello, 17th cent. *Letter.* 1633. 1 item (1 leaf)

A.L.S. (1633 May 7, Florence) to Galileo; in Italian.   MSS 1207 A

58. Galilei, Galileo, 1564-1642. *Letter.* 1635. 1 item (2 p.)

A.L.S. (1635 May 12) to Peiresc; in Italian. [v]   MSS 562 A

59. Rohault, Jacques, 1618-1672. *[De l'aymant].* 1638-1672. 1 item (7 p.)

A. ms. (undated) on magnets in 83 numbered statements, perhaps an outline or table of contents, concluding with a note in a contemporary hand: "Cet ordre d'experience de l'aymon est de la main de Monsieur Rohault"; in French.   MSS 1229 A

60. Kircher, Athanasius, 1602-1680. *Letter.* 1640. 1 item (1 leaf)

A.L.S. (1640 Dec. 28, Rome) in Latin.   MSS 783 A

61. Trew, Abdias, 1597-1669. *Autograph signature.* 1641-1647. 1 item (2 p.)

Probably a leaf from an autograph album: Trew's autograph signature (1641 Aug. [12/13?]) with Latin motto on recto, 1 autograph signature of Johannes Latermann (1647 Jan. 28) with Latin note on verso.   MSS 1481 A

62. Vossius, Gerardus Joannes, 1577-1649. *Letter.* 1643. 1 item (1 leaf)

A.L.S. (1643 Sept. 7, Amsterdam) to Grotius concerning fasting; in Latin.   MSS 1517 A

63. Guericke, Otto von, 1602-1686. *Letters.* 1647-1671. 3 items.

Three A.L.S.; in German, French, and Latin.   MSS 630 A

64. Zeisold, Johann, 1599-1667. *Papers.* 1649-1659. 3 items.

Holograph leaf, possibly an A.L.S. (1649 Dec. 7) in German, and two holograph leaves S. ([1]652 Jan. 29 and 1659 Apr. 6) from autograph albums, each containing the same Christian epigram in Latin.   MSS 1595 A

65. *Beschreibung von den gantzen Fundament was ein Feuerwercks Meister lehren und verstehen mus, und in summa, was zu der gantzen Artolleria gehorig sey.* [1650-1660]. 138 leaves [i.e. 304 p.]: ill.; 11 x 18 cm.

Inscription on page 1: Ex libris Jacobi Josephi Comitis in Wolkenstein Ao 1698. On spine: Artolleria - Feverwercks Meister - ca. 1650.   MSS 274 B

66.   *In physica.* [ca. 1700]. [1059] p.: ill.; 19 cm.

Holograph ms. on natural philosophy; in Latin. On spine: Physica, c. 1700.   MSS 240 B

67.   *Tractatus de experimentis mathematicis; [Brevis consideratio aliquorum problematum maxime nominatorum apud geometras]* [ca. 1700]. [125] p.: ill.; 21 cm.

Bound ms. (undated) with two treatises, two sections of "Problemata diversa," one of "Theoremata," and several pages of notes. The first treatise concerns "Experimenta bellica" and "Experimenta astronomica." The second treatise concerns "Circuli quadratura," "Cubi duplicatio," and "Anguli rectilinei trisectio"; in Latin. Spine title: Tract. Exper. Math.   MSS 1295 B

68.   **Vauban, Sébastien, Le Prestre, de, 1633-1707.** *[Traitté des siéges et de l'attaque des places]; [Traitté de la deffense(!) des places].* [ca. 1700]. [230] p., 33 leaves of plates: ill. (chiefly col.); 45 cm.

Holograph ms. S. containing dedicatory letter to Monseigneur Le Duc de Bourgogne and "cet ouvrage," namely the two treatises. The first has 25 chapters, indexed on pages 198-200, the second treatise has 5 chapters. Later published as *De l'attaque et de la defense des places;* in French. Spine title: Fortifications. [VI]   MSS 1303 B

69.   **Valentinus, Basilius, fl. 1413.** *Traité du grand oeuvre, ou, Troisième partie du testament de Frère Basile Valentin.* [1651-1699]. 100 p.

Holograph ms. on alchemy; in French. Pages 1-2 replaced from copy in Bibliothèque de l'Arsenal by previous owner, Dr. Marc Haven. Notes on slips used to mark pages in front. On spine: Basile Valentin-Alchemica-MS.   MSS 213 B

70.   **Dumas, Bernard, 17th cent.** *Breve compendium philosophiae Bernardi Dumas.* 1652. 364 p.: ill.; 15 cm.

Holograph ms. on natural philosophy; in Latin. Binding by Bernasconi. Spine title: Compendium philosophiae.   MSS 209 B

71.   **Nederven, Cornelius Van, 17th cent.** *Annotata physicalia.* 1657-1658. [472] leaves: ill. (some col.), map; 22 cm.

Treatises on Aristotle's *Physica* and *Metaphysica.* Also, under the title *Annotata minus principalia,* treatises on seven works including the *Sphaera* of Sacro Bosco, and minor works and poems (one partly in Dutch); in Latin. Partially rubricated.Second leaf: Scribebat Cornelius Van Nederven Braedanus Anno 1657. Spine title: Physica / Van Nederven. Back cover: Braedanus Anno 1658.   MSS 986 B

72.   **Gregory, James, 1638-1675.** *Calculations.* [between 1658 and 1675]. 1 item (1 p.)

Holograph page of calculations, undated.   MSS 619 A

73.   **Huygens, Christiaan, 1629-1695.** *Letter.* 1659. 1 item (2 p.)

A.L.S. (1659 Sept. 25, the Hague) to an unidentified correspondent sending some examples concerning the new chronology; in French.   MSS 739 A

74.   **Heweliusz, Jan, 1611-1687.** *Papers.* 1660-1680. 2 items: ill.

Two A. papers S. (1660 Nov. 9, and 1680 May 31); in Latin.   MSS 699 A

75.   **Newton, Isaac, Sir, 1642-1727.** *Chemical notes.* [1660-1727?]. 2 items.

A chemical recipe using gold and silver. Also, a second recipe beginning in Latin ("Materia sublimatur") on distillation, in which Newton cites Basilius. Burndy ms. no. 12.   MSS 1007 B

76.   **Newton, Isaac, Sir, 1642-1727.** *Ex Fabri hydrographo spagyrico; Ex Palladio spagyrico.* [between 1660 and 1727]. [4] p.; 30 cm.

Holograph notes or extracts from two alchemical texts, the first on the Fons Chemicorum and aqua vitae. Newton cites the page numbers of the editions he uses; in Latin. Burndy ms. no. 13.   MSS 1024 B

77. Newton, Isaac, Sir, 1642-1727. *Notanda chymica.* [between 1660 and 1727]. [5] p.; 24 cm.

Holograph notes and extracts from Maier's *Arcana arcanissima,* citing page numbers of edition; in Latin. Burndy ms. no. 14. Dealer's description laid in.

MSS 1028 B

78. Newton, Isaac, Sir, 1642-1727. *The Regimen.* [between 1660 and 1727]. 4 items; 23 cm.

Two similar texts, both called the Regimen, on a decoction outlined variously by Philaletha, Pontanus, and others. Also, "Of the Regimen," which begins: "By continual decoction thou shalt see in 130 days the white Dove and in 90 days more the sparkling Cherubim." Draws on Ripley and others. With fragment of a page (1689 Sept. 11) in Latin, citing Roger Bacon's *Elementorum* and Michael Maier's *Arcana arcanissima* and mentioning John Day. Partial transcription laid in. Burndy ms. no. 15.

MSS 1032 B

79. Newton, Isaac, Sir, 1642-1727. *Separatio elementorum; Reductio et sublimatio.* [between 1660 and 1727]. 2 items; 34 cm.

Newton's Separatio elementorum (2 p.) outlines the distillation of elements in a 20 gallon vat and cites Philaletha and Magnus. Reductio et sublimatio (5 p.), extracted from the work of Raymond [Lull?], on the imperfect transmutation of white and red sulphur into silver and gold respectively; in Latin. Spine title: Separatio / et / Reductio. Burndy ms. no. 10.

MSS 1041 B

80. Newton, Isaac, Sir, 1642-1727. *Vegetation of metals.* [between 1660 and 1727]. [12] p.; 22 cm.

Holograph notes on vegetation and the generation of minerals, as well as air, heat, fire, cold and God. Last page only in Latin. Spine title. Burndy ms. no. 16. [VII]

MSS 1031 B

81. Redi, Francesco, 1626-1698. *Letter.* 1660. 1 item (2 p.)

A.L.S. (1660 Oct. 6, Florence) to Carlo di Dottori, Padua, concerning "l'acqua argente bianca" and "l'acqua colorita" and mentioning the recipient's two canzone; in Italian.

MSS 1195 A

82. Arnaldus de Villanova, d. 1311. *Mag-[istr]i Arn[al]di Devillan[ov]a Liber dictus Nouum Lumen; item Rayme[n]di Lullij Potestas Diuitatia[rum] Lapidarium & Ultimu[m] Testam[entu]m.* 1666. [3], 402 p.: ill.; 13 cm.

MSS 272 B

83. Boyle, Robert, 1627-1691. *Letter.* 1666. 1 item (4 p.)

A.L.S. (1665/6 Jan. 18, Oxford) to the Earl of Burlington requesting the appointment of some trusty person to receive Boyle's Irish rents.

MSS 161 A

84. *Delle prepar[azio]ni de metalli et del modo d'estrahere le pure essenze dell'oro chimistichamente.* [late 17th century]. 79 leaves [i.e. 154 p.]: ill.; 19 cm.

Half sheet laid in at leaf 34. Spine title: Alchimia M.S. Dealer note: Termina con ricette di Paracelso.

MSS 218 B

85. Cassini, Giovanni Domenico, 1625-1712. *Letters.* 1671-1708. 2 items.

Two A.L.S. (1671 Dec. 17 and 1708 Nov. 26, Paris), the latter thanking the secretary of the Royal Society of London for a letter of condolence on the death of Mme. Cassini; in French.

MSS 310 A

86. Wallis, John, 1616-1703. *Letter.* 1671. 1 item (2 p.)

A.L.S. (1671 June 2, London) to an unnamed correspondent on p. 44 removed from a book, in Latin. Wormholes in leaf.

MSS 1529 A

87. *Magneticall observations.* 1673-1674. 54 [i.e. 53] leaves: ill.; 30 cm.

Notebook on the properties of the lodestone. First entry dated "20 Martii 1673"; last dated entry is "7 Sept. 1674." Additional materials includes English loadstone with iron armor and keeper and description provided by Burndy Library. Spine title: MS of magnet; box title: Lodestone with magnet.

MSS 938 B

88.　Ward, Seth, bishop of Salisbury, 1617-1689. *Document.* 1673. 1 item (1 leaf)

A.D.S. (1673 Sept. 25, Chute Lodge, Wells) to the Churchwardens of the parish of Denchworth assigning the Curé of Denchworth to William Milner.

MSS 1532 A

89. *Geometriae rudimenta in gratiam eorum qui philo[so]phiae operam sunt daturi.* [ca. 1700.] 53 p., [14] folded leaves of plates: ill.; 23 cm.

MSS 231 B

90.　Merles, Balthazard-François de, 17th cent. *Papers.* [4th quarter of 17th century]. 13 items: ill.; 31 cm.

One A.L.S. (1699 Mar. 20, Marseille) in French to de Beauchamp at Avignon on the recent eclipse; preserved in same vellum folder with 3 other items: notes on solar eclipse (1696 Sept. 20), notes on a total eclipse (1693 Jan. 21), and a diagram of an eclipse. Six other folders, five containing notebooks: "De epochis seu aevis" (with table on days of the month) in Latin; how to "dresser une figure caeleste" by several methods, in French; and other astronomical topics. One folder contains 2 lunar maps and a diagram of a lunar eclipse in ink and wash.　MSS 859 B

91. *Physica seu na[tura]lis philosophia.* [late 17th century]. [466] p.: ill., one col.; 20 cm.

Contents: Pars prior, physica generalis, sectio I. De materia--II. De forma--III. De motu--IV. De qualitatibus--Altera pars, physica specialis, sectio I. De mundo celesti--II. De mundo sublunari--III. De plantis et animalibus--IV. De homine; in Latin.　MSS 1252 B

92. *Mercurii zweyfacher Schlangen-Stab.* 1679. [4], 124 p.: ill.; 17 cm.

Subtitles: I. Glücks-Ruthe zu Paracelsi chymischem Schatz. II. Menstruum oder philosophischen Solvens universale. . . samt dem gantzen philosophischen Process. Possibly a transcription of the 1678 edition (subtitles vary slightly). The two-part alchemical work draws on Paracelsus. Page [4], the heading page for "Glücks-Ruthe," has the date 1627; in German. With a tractate by Nuysement.

MSS 939 B

93.　Montanari, Geminiano, 1633-1687. *Letter.* 1679. 1 item (2 p.)

A.L.S. (1679 Aug. 24, Venice); in Italian.

MSS 1029 A

94.　Kunckel, Johann, von Löwenstern, 1630?-1703. *Documents.* 1680. 2 items.

Two A.D.S. (1680 Aug. 20, Berlin, and 1680 Aug. 25); in German.　MSS 805 A

95.　Mariotte, Edme, ca. 1620-1684. *A treatise of the motion of water and other fluid bodyes.* [1690?]. 123 p.: ill.; 18 cm.

English translation of Mariotte's "Traité du mouvement des eaux et des autres corps fluides." Preserved in box with spine title: Motion of Water (1690).

MSS 908 B

96. *Mechanicks; Opticks; Astronomy.* [before 1729]. 401 leaves: ill.; 21 cm.

Spine title: Manuscript on Mechanics. Notes on front flyleaf: "E : Libris Chris. Burton Divi : Joh : Coll : Cant : Alum :" and "1729 / Paul Burton."　MSS 964 B

97.　Placentinus, Jacobus, 1673-1762. *De percussione et legibus motus corporum percussorum libellus auctore viro doctissimo clarissimoque Jacobo Placentino in Patavina Universitate Libri tertii Avicennae interprete.* 1680. [51] leaves: ill.; 22 cm.

Ms. written on versos only; in Latin. Spine: Placentino--De percussione--1680.　MSS 1254 B

98.　Louys, Sebastianus David, 17th cent. *Gheometria oft de konste van Landt te meten.* 1681. 294 [i.e. 296], [76] p.: ill. (some col.); 21 cm.

On plane and solid geometry, land measurement, and triangulation, primarily. Rubricated, 6 fold-out illustrations. Margins drawn & numbers marked on pages 1-244, the main text. Minor texts added on following unnumbered pages; including a Latin poem and epigrams and a two-page problem in French; in Dutch.　MSS 827 B

II.   Drawing of military machinery, possibly for Ramelli's
      Le diverse figure et artificiose machine (1588).
      Entry No. 11, SI Neg. 84-7170.

99. Morland, Samuel, Sir, 1625-1695. *A Memorial concerning ye mill and engin at Windsor.* 1685. 1 item (9 p.): ill.

"Given into His Majesty's own hands 2 Mar. 1685." MSS 1039 A

100. Clerke, Gilbert, 1626-1697? *Correspondence.* 1687. 4 items.

Four A.L.S. (1687 Sept. 26-Nov. 21, St. Martins) from Clerke to Newton on difficult passages in Newton's *Principia.* On the first, Newton wrote an A.L.S. to Clerke in reply. Burndy ms. no. 11. MSS 1008 B

101. *De la gnomonique, ou, La Manière de faire des cadrans.* [ca. 1700]. 47 p.: ill.; 17 cm.

Spine title: Gnomon. Dealer description laid in. MSS 220 B

102. *De natura ac physices obiecto.* [169-?]. [512] p.: ill.; 19 cm.

Date on spine. MSS 238 B

103. Newton, Isaac, Sir, 1642-1727. *Chemistry transcriptions.* [between 1693 and 1727]. [3], 61 p.; 33 cm.

Holograph transcriptions and abstracts of chemical texts including five treatises from Villanova's *Rosarium abbreviatum,* Boni's *Margarita Pretiosa,* and Philaletha's English texts on sulphur. Two texts are from Blankaart's *Theatricum chimicum* (1693). Ms. ends with 208 blank pages. Note on first flyleaf: "Sept. 25, 1727 / Not fit to be printed / Tho. Pellet." In Latin. Spine title: Chemistry Manuscript. Annotated table of contents laid in. Burndy ms. no. 9 MSS 1023 B

104. Duhan, Laurent, 1656-1726. *Compendiaria ad universam ph[ilosoph]iam introductio dictata a domine Laurentio Duhant.* 1696. 227, [38] p.: ill.; 17 cm.

Ill. on back lining paper. Index. With part of *Prima disputationum rumdimenta, in gratiam, adolescentium ad res philosophicas contendentium,* a printed work (32 p.). MSS 217 B

105. Flamsteed, John, 1646-1719. *Documents.* 1696-1718. 3 items.

Three A.D.S. certifying how long and at what pay three men worked for the observatory. MSS 518 A

106. Keill, John, 1671-1721. *An examination of Dr. Burnet's Theory of the earth; Some remarks on Whiston's [New] theory of the earth.* 1698-[1745?]. 329, [35] p.: ill.; 21 cm.

Published by Keill at Oxford, "Printed at the Theatre," 1698; in English with Greek closing. Included in the same codex are "Mr. Saunderson's lecture upon Mechanicks, Astronomy, Opticks & Hydrostaticks"; "Hydrostaticae praelectiones," in Latin; "Doctrine of sound," "Doctrine of heat & cold," "An account of the Rainbow"; "An addition to the foregoing astronomy"; "Ode in poem," in Latin; and "A Catalogue." Three catalogue entries are dated: 1736-1745. MSS 840 B

107. Lichtscheid, Ferdinand Helffreich, 1661-1707. *Letter.* 1699. 1 item (1 leaf)

A.L.S. (1699 Aug. 4) to be sent with 24 copies of his book and dealing with the intended purchase of certain publications; in German. MSS 894 A

108. *Abbregé de physique.* [18th century]. 119, [21] leaves.

On spine: Physique W.C. With: Theses physicae (21 p.) in Latin. MSS 24 B

109. *[Aritmetica].* [early 18th century]. [263] p.; 29 cm.

On spine: Aritmetica-1700. Various notes laid in or tipped in. MSS 262 B

110. *Cartas de sphera caelesti juxta miram angelici doctoris, et arlis doctrinam.* [18th century]. 226 p.: ill.; 24 cm.

With: Tractatus de generatioe[sic] et corruptione. MSS 211 B

111. Colomboni, Angelo Maria, 1608-1672. *Arte gnomonica del padre D. Angelo Maria Colomboni.* [18th century]. 52, [39] p.: ill. (some col.); 31 cm.

Many blank leaves in center of volume. On spine:
Colomboni Arte Gnomonica Figure Mss. 186.

MSS 257 B

**112. Columella, Lucius Junius Moderatus,
fl. 60.** *Delle cose de la villa libro secondo; Libro
degl'Abberi.* [18th century]. 179 leaves [i.e. 358
p.]; 25 cm.

Italian translations of parts of the first century works
De Re Rustica and *De Arboribus* of Columella. On end
leaf: Ad usum Petri Vincj J.C. Florentinj. MSS 278 B

113. *De alchimia tractatus.* [18th century].
[1], [184] p.

Title ([1] p.) in different hand. MSS 7 B

114. *Della sfera.* [18th century]. 28 p.; 30 cm.

Holograph ms. on the sphere; in Italian. MSS 281 B

115. Dolce, Lodovico, 1508-1568. *Delle
gemme che produce la natura della qualita.*
[18th century?]. [150] p.; 17 cm.

With an incomplete treatise on the significance of im-
ages carved on gems. Slip-case title: Delle Gemme.

MSS 208 B

116. *Fabbrica et uso del compasso di quattro
punte.* [18th century]. 94, [10] p.: ill.; 28 cm.

Spine title: Compasso di quattro punte. MSS 260 B

117. Glinigh, Giovanni Girolamo. 18th cent.
*Un modo breve & facili per far gl'horologii ori-
zontali italiani per via di numeri del P. Gio.
Girolamo Glinigh della Comagnia d'Gresu,
scriferito da D. Gio. Francesco Palmieri senese.*
[18th century]. 10, [8] p.: col. ill.; 20 cm.

MSS 223 B

**118. Kramer, Johann Georg Heinrich, d.
1742.** *Letter.* [between 1700 and 1742]. 1 item (2 p.)

A.L.S. (undated) to D. Jacquin (?); in Latin.

MSS 796 A

119. Lavater, Jean-Gaspard, 1741-1801. *Envelope.* [18th century?]. 1 item (1 leaf)

A. envelope addressed to Madame Meynel; in French.

MSS 856 A

120. *[Lectures on natural philosophy].* [18th
century]. 106 [i.e. 107] leaves: ill.; 21 cm.

Twenty-one lectures on natural philosophy. Topics include mechanics, hydrostatics and pneumatics, optics,
and Newton's theory of light and colors. Outer margin
ruled, with a few marginal diagrams and several references
to plates and figures not in the ms. Text followed by 66
blank leaves and then "A catalogue of experiments con-
teyned [sic] in this book." MSS 833 B

121. Lucretius Carus, Titus, 94-55 B.C. *T.
Lucrezio Caro Della natura delle cose libri VI
tradotti in versi sciolti da Alessandro Marchetti
Pistojese.* [between 1700 and 1714]. 233 leaves; 31
cm.

Marchetti's Italian translation of *De Natura rerum*.
Written on recto and verso, numbered on recto only.
Table of contents follows text. Title page notes that Mar-
chetti is known among the Arcadians as "Alterio Eleo."

MSS 829 B

122. Newton, Isaac, Sir, 1642-1727. *Papers.*
1700-1718. 5 items.

A. ms. of 42 p. on alchemy, drawing from several
works, especially those of Llull (Raymond Lull) and Ripley; one "opus" is on the "Regimen decoctionis"; in Latin.
Three A.L.S. (1700, 1718, n.d.) and a transcript of Bernoulli's letter (1717) to Montmort about Newton and
Leibniz, annotated by Newton; in English.

MSS 1070 A

123. *Les principes de la sphere; De la carte ou
mappe universale.* [18th century]. 101 [i.e., 202] p.:
ill.; 19 cm.

Manuscript on astronomy and geography, especially of
France, Italy, and Germany, with 154 blank pages after
text; in French. Spine title: Sphere. MSS 1258 B

124. *Drawing of a pump.* [18th century].
1 item (1 leaf): col. ill.

Folded pen and wash illustration of a pump for raising
water, labeled "fig. 3," with labels in French.

MSS 1179 A

125. Saunderson, Nicholas, 1682-1739. *[Mathematical treatise].* [18th century]. 149 leaves: ill., port.; 21 cm.

Manuscript on several mathematical topics including the algorithm of fluxions, the mensuration of solids, the quadrature of curves, and forces; possibly an early draft of his *Method of Fluxions;* 20 blank leaves at end. Engraved portrait from *Gent[leman's] Mag[azine]* of Sept. 1754 pasted in. MSS 1276 B

126. *Tavola del meriggio calcolata per l'elevazione del polo di gradi 413, minuti 51 latitudine, e di gradi 46 e minuti 9 dell'altezza dell'equatore.* [18th century?]. 1 item (1 leaf): col. ill.

Rubricated ms. astronomical calendar table showing midday for each day of the year. MSS 1457 A

127. *Traité de méchanique.* [18th century?]. 67 [i.e., 135] p.: col. ill.; 27 cm.

Bound ms. containing one text in four parts: "Du mouvement des corps durs sans ressorts," "Des corps a ressorts," "Du mouvement des corps pesants," and "Des machines," includes 24 folded plates of pen and wash drawings; in French. MSS 1297 B

128. *Trattato d'aritmetica.* [ca. 1750]. lxvii [i.e., 48] p.: ill.; 17 cm.

Bound ms. for a child to learn arithmetic, including pen and ink drawings of flowers, birds, and insects. "Raffaello Beccattini" written on end flyleaf; in Italian. Page xxiii/xxiv torn out, pages xxxix-liiii lacking. Spine title. MSS 1299 B

129. *Trattato di gnomonica ovvero regole per fare l'orologj a sole, ec.* [18th or 19th century]. 1 item (17 p.): ill.

Manuscript on telling time by the sun (15 p.) and telling time by the moon (1 p.). The text notes "Weitlerio pag. 281" as the source of the information. The last page has a note on "Duplications del Cubo"; in Italian. MSS 1480 A

130. Christian, duke of Saxe-Eisenberg, 1653-1707. *Letter.* 1703. 1 item (1 p.)

A.L.S. (1703 Dec. 20) sending New Year's wishes to his aunt and sister-in-law; in German. MSS 343 A

131. Leibniz, Gottfried Wilhelm, Freiherr von, 1646-1716. *Letters.* 1703-1714. 4 items.

Four A.L.S. to various correspondents including Fontenelle; in French and Latin. MSS 875 A

132. Carapecchia, Romano Fortunato, 17th-18th cent. *La forza della leva opera di Romano Fortunato Carapecchia.* 1704. [9] p., 73 leaves [i.e. 145 p.]: ill. (one folded); 25 cm. MSS 245 B

133. Copley, Godfrey, Sir, d. 1709. *Letter.* 1704. 1 item (2 p.)

A.L.S. (1704 April 29) to Thomas Kirk concerning an air pump and a loadstone. MSS 373 A

134. Manfredi, Eustachio, 1674-1739. *Letters.* 1704-1734. 2 items.

A.L.S. (1704 Oct. 22, Bologna) in French, and A.L.S. (1734 May 10, Bologna) in Italian, both to Maraldi. MSS 953 A

135. Fontenelle, Bernard Le Bovier de, 1657-1757. *Document.* 1705. 1 item (1 p.)

D.S. (1705 July 2, Paris) as receipt for a payment; in French. MSS 525 A

136. Roques, Pierre, 1685-1748. *Cours de physique composé par Monsieur Pierre Roques.* [between 1705 and 1748]. 914, [7] p.; 24 cm.

Manuscript in 4 separately numbered parts, each with a table of contents: I. [Des principales qualités communes à tous les corps]--II. De la cosmographia--III. De la terre--IV. [Des corps animés]; in French. Spine title: Physique. MSS 1271 B

137. *[Mercantile arithmetic].* 1707. 115 leaves: ill. (some col.); 20 cm.

On arithmetic, esp. concerning weight, distance, time, and money; on allegation. Partially rubricated. On leaf 67 "this present year of our Lord" is 1707. Spine title: Arithemtick / 1707. MSS 965 B

138. Keill, John, 1671-1721. *Letter.* 1709. 1 item (8 p.)

A.L.S. (1709 Dec. 5) to Wolff; in Latin.   MSS 775 A

139. Hamberger, Georg Albrecht, 1622-1716. *Autograph signature.* 1710. 1 item (1 p.)

Autograph signature (1710 May 8, Jena) with Biblical quotation; in Latin and Greek.   MSS 657 A

140. Lagomarsino, Lorenzo, 17th-18th cent. *Geografia. . . Cosmographia. . . Gnomonice.* 1711-1713. [242] p.: ill.; 21 cm.

Three treatises (1711-1713), each illustrated with diagrams and indexed. Treatise on sundial has diagram with moving dial. With Philip Palavicinus' Questiuncules miscelanes(!) (1712) and the De Fulmine (1713) of Antonius de Magistris. Owner's note in Italian on p. 1 of first text. Coat of arms containing "D.F." stenciled on 4 pages. Signed on p. 1: Dominicus Maria Fornelli; in Latin.   MSS 831 B

141. Sturm, Leonhard Christoph, 1669-1719. *Letter.* 1714. 1 item (1 leaf)

A.L.S. (1714 Jan. 26, Schwerin); in German.   MSS 1433 A

142. Voltaire, François Marie Arouet de, 1694-1778. *Papers.* [between 1714 and 1778]. 2 items.

Holograph ms. (2 p.) concerning "Relligion"(!) and containing a date (1754 June 12) in one entry, and an incomplete A.L. to "Gabriel Crammer"; in French.   MSS 1516 A

143. Böttger, Johann Friedrich, 1682-1719. *Document.* 1716. 1 item (1 p.)

A.D.S. (1716 Dec. 14); in German.   MSS 130 A

144. Wolff, Christian, Freiherr von, 1679-1754. *Letters.* 1716-1746. 3 items.

One A.L.S. (1716 May 1, Halle) in French and German to Barth concerning Hauksbee's experiment, and two A.L.S. (1744 Feb. 26 and 1746 Oct. 22, Halle), the former concerning sacred theology; in German.   MSS 1583 A

145. Gornia, Victorio, 17th-18th cent. *Dissertatio medico-experimentalis que, coram eminentissimo et reverendissimo D.D. Curtio Orighi bononie a latere legato, recitanda erat die 18 Mensii [?---bry], anni 1717, in Publica Academia Instituti Scientarum Bononie a Victorio Gornia.* 1717. [21] p.; 26 cm.   MSS 228 B

146. Lucretius Carus, Titus, 94-55 B.C. *Della natura delle cose libri sei.* [after 1717]. [19], 432, [10] p.; 23 cm.

Marchetti's Italian translation of *De natura rerum*, copied from a printed edition; title page included with publication details: prima edizione, Londra a Giovanni Pickard MDCCXVII [i.e. London: John Pickard, 1717]. Name in another hand on the last leaf.   MSS 828 B

147. Réaumur, René-Antoine Ferchault de, 1683-1757. *Letter.* 1717. 1 item (4 p.)

A.L.S. (1717 Jan. 10, Paris) to Monsieur de Jussieu comparing charcoal ("le charbon des bois") and coal ("le charbon de terre"); in French.   MSS 1194 A

148. Maupertuis, Pierre Louis Moreau de, 1698-1759. *Letters.* [between 1718 and 1759]. 2 items.

Two A.L.S. (undated), one supposedly to Voltaire; in French.   MSS 976 A

149. Montmort, Pierre Rémond de, 1678-1719. *Letters.* 1718. 2 items.

A.L.S. (1718 Mar. 27) to Newton concerning the problem of trajectories set by the Bernouillis, in French, Latin, and English, and A.L. (1718 Dec. 18) to Taylor concerning other scientists that Taylor should know, in French.   MSS 1035 A

150. Sorra, Francisco Maria, 17th-18th cent. *Physicae institutiones a Francisco Maria Sorra Mutinensi scriptae.* 1718. [412] p.: ill. (one col.) 23 cm.

Bound ms. with three parts: "Physica generalis" (in 5 chapters, the longest concerning motion), "Tractatus de caelo et mundo" (in 5 chapters), and "Tractatus de quattuor corporibus"; diagrams; in Latin. Spine title: Physicae. One leaf laid in. MSS 1283 B

151. Varignon, Pierre, 1654-1722. *Letters.* 1718-1719. 2 items.

A.L.S. ([1718] July 22) to Birr requesting a book for Bernoulli; and A.L.S. ([1719] Aug. 30) to Westein concerning a pacquet for Bernoulli; in French.

MSS 1500 A

152. Frisi, Paolo, 1728-1784. *Letters.* 1722. 2 items.

A.L.S. (1772 Dec. 2, Milan) and A.L.S. (1772 Dec. 22, Milan); in Italian. MSS 550 A

153. Bernoulli, Jean, 1667-1748. *Letters.* 1723-1728. 2 items.

Two A.L.S. to Mairan and Cramer; in French.

MSS 96 A

154. Payson, Phillips, 1705-1778. *A commonplace book.* 1723-1726. 170 [i.e., 182] p.; 16 cm.

Holograph which treats causes and atoms briefly and theology at length. Fly leaf has author's signature three times with the dates "1723/24" and "1726." Spine Title: Common Place Book. MSS 1249 B

155. Stearne, Robert, 18th cent. *A collection of geographical, astronomical, and astrological problems corrected from the observations communicated to the Royal Society's of London and Paris; also the Theory of the tides from Sir Isaac Newton's Works; likewise an attempt to assign the physical cause of the trade winds and monsoons by Doct[o]r Edm[on]d Halley.* 1723-1730. Holograph signed (1730) and written in 1723 and 1724 (cf. p. 491). Topics include the work of Flamsteed, Halley, Newton and Whiston, figuring the golden number and the domenical letter, calendars, the eclipses of 1715 and of 1725-40, and major rivers. Two pages in Latin. Cover title: A Collection of Geometrical(!) & Astronomical Problems.

"Collected at his Majesties Royal For[t] of Duncannon Decemb[e]r 1723. By the Honourable Brigad[ie]r Gen [era]l Rob[er]t Stearne and Gov[erno]r of the Said Fort."

MSS 1288 B

156. Bernoulli, Daniel, 1700-1782. *Letters.* 1724-1768. 2 items.

Two A.L.S., one in Latin with mathematical formulae, and one in French to Mairan. MSS 95 A

157. Doppelmayr, Johann Gabriel, 1671-1750. *Letters.* 1724-1730. 4 items.

Four A.L.S. to Liebknecht; in German. MSS 449 A

158. *[De physica].* [mid-18th century?]. 2 v.: ill.; 21 cm.

Spine title: Fisica. MSS 265 B

159. *Della sfera armillare trattato che serve di introduzione alla geografia.* [ca. 1750]. 119 leaves [i.e. 238 p.]: ill.; 21 cm. MSS 239 B

160. Desaguliers, John Theophilus, 1683-1744. *Letter.* 1725. 1 item (3 p.)

A.L.S. (1725 April 29) to the president of the Royal Society (Newton) concerning experiments which Desaguliers made at each meeting. MSS 430 A

161. Euclid, fl. ca. 300 B.C. *Questo libro et di Francesco di Baccio di Francesco Saminiati nel quale scriverra tutti li propositioni d'Eucrido [sic].* [mid-18th century]. [47] p.: ill.; 30 cm.

MSS 252 B

162. Guinée. *Traité de mechanique de M. Guinée.* [ca. 1735]. 48 leaves [i.e. 96 p.]: ill. (some folded); 26 cm. MSS 227 B

163. Boerhaave, Herman, 1668-1738. *Letter.* 1728. 1 item (4 p.)

A.L.S. (1728 May 4, Leyden) to Miller concerning plants which Miller had sent; in French. MSS 129 A

164.   Hoffman, Friedrich, 1660-1742. *Epigram.* 1728. 1 item (1 p.)

A. epigram S. (1728 Mar. 24, Halle) on essential traits of a good man, philosopher, and judge; in Latin. Annotated on verso by another hand with biographical information about Hoffman.   MSS 709 A

165.   Lehmann, Johann Christian, 1675-1739. *Letter.* 1728. 1 item (4 p.)

A.L.S. (1728 Feb. 8, Leipzig) requesting a decision regarding his complaint in connection with the threatened removal of his graduation barrels; in German.

MSS 873 A

166.   Sloane, Hans, Sir, 1660-1753. *Letter.* 1728. 1 item (2 leaves)

A.L.S. (1728 Dec. 3) concerning books and recent astronomical discoveries by Bradley and Halley; in French.
MSS 1378 A

167.   Belidor, Bernard Forest de, 1697?-1761. *Traité ou mémoire sur une nouvelle théorie de la science des mines par Belidor.* 1729. 32 p.; 38 cm.

Cover title.   MSS 287 B

168.   Grandi, Guido, 1671-1742. *Letter.* 1729. 1 item (6 p.)

A.L.S. (1729 Feb. 11, Pisa) to Diego Sanduccio chiefly on sixteenth-century history of regional places, includes a lengthy quotation in Latin from a 1575 text. Evidence of a seal. The acidic ink has damaged the paper, now encapsulated in mylar; in Italian.   MSS 851 A

169.   Grandi, Guido, 1671-1742. *Papers.* 1729-1742. 16 items.

Twelve A.L.S. to various correspondents, 2 A.L.S. in a different hand, and 2 sheets of notes on the inscription on Grandi's tombstone; in Italian.   MSS 612 A

170.   Dutour, Etienne François, 1711-1784. *Correspondence.* 1730-1770. 148 items: ill.; 28 cm.

Correspondence between Nollet, French abbé and physicist, and Dutour, theologian and natural philosopher, relating to discoveries and scientific quarrels of the period. The collection includes letters from Defouchy (1746), Noël Delor (1751), Cavelier (1756), and Chupey

(1767) to Dutour. Spine title: Nollet / Dutour / Correspondence / 1742-1770. Dealer's description pasted in.
MSS 1061 B

171.   Nollet, Jean Antoine, 1700-1770. *Marginalia.* 1731-1784. 8 items: ill., maps; 28 cm.

Nollet annotated in French the first book of Beccaria's Italian text *Dell'elettricismo artificiale e naturale* (1753); the marginalia was truncated when the text was bound with 6 other printed scientific texts, including Nollet's works on icebergs and the transmission of sound through water (1743), Bouguer and Bazin on the compass (1731 and 1753, the latter with 15 plates by Striedbeck), Desaguliers on electricity (1742) and Dutour on magnetism (1746). Dutour annotated his title page in 1748; this volume bears his bookplate. His title on flyleaf: Pièces sur le magnetisme et electricité; in French. Transcription of some marginalia laid in. Spine title: Recoei(!) de pieces.
MSS 1060 B

172.   Aguesseau, Henri François d', 1668-1751. *Letters.* 1732. 2 items.

L.S. (1732 Oct. 20) with 3 autograph lines from Aguesseau to M. de Collarès, président au Conseil supérieur à Perpignan; A.L.S. (undated) to unidentified; in French.
MSS 20 A

173.   Bertin, Henri Léonard Jean Baptiste, 1719-1792. *Papers.* 1732-1760. 2 items.

D.S. (1732 March, Paris) acknowledging receipt of tax; A.L.S. (1760 March 5, Versailles) acknowledging receipt of letter; in French.   MSS 104 A

174.   Euler, Leonhard, 1707-1783. *Papers.* 1732-1766. 5 items.

Four A.L.S. (in French and German), and 1 holograph article (in Latin), "L. Euleri Constructio aequationum quanundam differentialium quae indeterminatanum separationum non admittunt. . ."   MSS 490 A

175.   Diderot, Denis, 1713-1784. *Poem.* [between 1733 and 1784]. 1 item (1 leaf)

A. poem S., eleven lines of poetry about friendship; in French.   MSS 439 A

176. L'Isle, Joseph Nicolas de, 1688-1768.
Letters. 1733-1741. 2 items.

A.L.S. (1733 Feb. 6/17, Petersburg) concerning the placement of Russian rivers on M. Kemfer's map, and A.L.S. (1741 Oct. 14, Petersburg) to the professors of the Académie des Sciences concerning a controversy with Heinsius; in French.   MSS 552 A

177. Mairan, Dortous de, 1678-1771. *Letters.* 1733-1756. 3 items.

Three A.L.S. (1733 Oct. 28, 1755 Feb. 3, and 1756 Feb. 19, all Paris), one to Abeille mentioning Nollet; in French.   MSS 951 A

178. Hauksbee, Francis, 1688-1763. *Letter.* 1734. 1 item (1 p.)

A.L.S. (1734 Aug. 28) to Nourse(?) deferring an engagement and enclosing a note.   MSS 673 A

179. Giardino, Alphunsius, 18th cent.
Philosofiae scripta mei Joannis Ancini Regiensis lectore Patre Alphunsio Giardino ordinis S. Barnabe S.T.D. in publico Mutinensi lyceo. 1735. 606, [1] p.: ill. (some col., one folded); 22 cm.
                MSS 246 B

180. *Tratado quinto dela artilleria.* 1735. [8], 140 [i.e., 279] p.: ill.; 22 cm.

Bound ms. (1735 Mar. 8), includes 11 folded plates; in Spanish. Spine title: Artilleria.   MSS 1298 B

181. Watson, William, Sir, 1715-1787. *Papers.* [between 1735 and 1787]. 2 items.

A.L.S. concerning Herschel's conversation with the king on astronomy and music, and a holograph ms. S. ([1748 June 16]): "An account of a treatise of Wm. Brownrigg, M.D., F.R.S., intitled the Art of making common Salt, as now practised in most parts of world; with several improvements proposed in that art, for the use of the British dominions" (17 leaves).   MSS 1536 A

182. Reichel, Christophorus Augustus, 1715-1742. *Letter.* 1736. 1 item (2 p.)

A. note S. (dated twice: 1736 Sept. 24 and 1736 Oct. [lit., 86] 1, Altdorf) to or concerning Johannes Golke, quoting Greek and a verse from the *Aeneid* in Latin; in German.   MSS 1198 A

183. Baratier, Jean Philippe, 1721-1740. *Letter.* 1738. 1 item (3 p.)

A.L.S. (1738 Feb. 23, Halle) to the Académie roiale des sciences concerning Baratier's proposed method of finding longitude; in French.   MSS 45 A

184. Meier, Georg Friedrich, 1718-1777. *Papers.* [1758-1761]. 2 items.

A.S. (1758 Feb. 14, Halae) with Latin epigram, and A.D.S. (1761 Oct. 11, Halae); in Latin.   MSS 985 A

185. Cassini, Jacques, 1677-1756. *Documents.* 1739-1747. 4 items

Four A.D.S., including a recommendation of a compass, 2 expenditure lists, and a request for funds; in French.   MSS 311 A

186. Du Châtelet-Lomont, Gabrielle Emilie Le Tonnelier de Breteuil, marquise, 1706-1749. *Letter.* 1739. 1 item (3 p.)

A.L.S. (1739 Jan. 19, Cirey) to Trublet concerning her name being compromised in the press; in French.
                MSS 456 A

187. Maclaurin, Colin, 1698-1746. *Letter.* 1739. 1 item (2 p.)

A.L.S. (1739 April 14, Edinburgh); photocopies kept in folder.   MSS 941 A

188. Tajoli, Felix, 18th cent. *Aristotelis logicam quam secutus est et publicè Senis edocuit P. Felix Tajoli in Collegio S. Vigilii anno 1739 me audiente Caesare Duratio convictore Collegii Ptolaemei.* 1739. 115 [i.e., 228] p.: ill.; 22 cm.

Holograph ms. signed on verso of title page: Cesare Durazzo. Includes an engraving signed Gottifredo di Blessi; in Latin.   MSS 1291 B

189. Celsius, Anders, 1701-1744. *Document.* 1740. 1 item (1 p.)

D.S. (1740 Sept. 20, Upsala), torn; in Swedish.
                MSS 320 A

190. *Physica.* 1740. [288] p.: ill.; 23 cm.

In two parts, the first on physics with topics including motion, hydrostatics, and elements, and the second part on astronomy and calendars; each part has a colophon dated 1740; in Latin.   MSS 1251 B

191. Cassini de Thury, César François, 1714-1784. *Papers.* 1741-1756. 2 items.

A. note S. (1741 Mar. 3, Paris) also signed by Grandjean de Fouchy, recommending publication of a book, and A.D.S. (1756 July 14) also signed by Camus, authorizing payment for engravings; in French.   MSS 313 A

192. Louis XV, king of France, 1710-1774. *Document.* 1741. 1 item (1 leaf)

D.S. (1741 Feb. 28, Versailles) concerning the transfer of a prisoner; in French.   MSS 924 A

193. Trew, Christoph Jacob, 1695-1769. *Autograph signature.* 1741. 1 item (1 leaf)

Autograph signature (1741 Oct. 11, [Nuremburg?]) with Latin epigram.   MSS 1482 A

194. Duhamel du Monceau, M., 1700-1782. *Papers.* 1742-1746. 9 items; ill.

A.D.S. (1746 Feb. 12) with Bouguer in memory of Mandillo, A.D.S. by Basseposte written for Duhamel de Monceau and Jussieu, and 7 ms. scraps; in French.   MSS 459 A

195. Guédier de Saint Aubin, Henri-Michel, 1695-1742. *Philosophiae partis [terti]ae seu phisicae reliqua Saint Aubin.* 1742. p. [447]-1134, i-xxv: ill.; 17 cm.

Manuscript volume 3 of a larger work, Philosophia, containing the conclusion of "[Pars prima seu phisica generalis]" and the whole of "Pars secunda seu phisica particularis" of "Phisica," which is "pars tertia" of the larger work; in Latin. Spine title: Cursus Philosop-[hiae] Tom[us] 3.   MSS 1273 B

196. Soumille, Bernard-Laurent, abbé, d. 1774. *Mémoire de M. l'abbé Soumille pour la construction d'une boussolle pour observer l'inclinaison.* 1742. [12] p.: col. ill.; 25 cm.

Bound ms. draft of a "Programme" submitted to the Académie in competition for its prize in 1743. Text concludes with the note "envoyé le 21 Juin 1742 avec un billet cacheté contenant le nom de l'auteur." Title added on the flyleaf, perhaps by Duhamel; in French. Spine title: Boussole.   MSS 1284 B

197. Carpov, Jakob, 1699-1768. *Papers.* 1743-1760. 2 items.

Signed holograph card of epigrams (1743, Vinaria) and signed holograph testimonial for C. Kiesewetter (1760 April 18, Vinaria); in Latin.   MSS 305 A

198. Bouguer, Pierre, 1698-1758. *Document.* 1744. 1 item (1 p.)

A.D.S. (1744 June 3, Nantes) recommending Pierre Fouré as a very capable pilot; in French.   MSS 155 A

199. Kant, Immanuel, 1724-1804. *Papers.* [between 1744 and 1804]. 3 items.

Three pages of autograph notes, none dated or signed; in German.   MSS 767 A

200. Manfredi, Gabriello, 1681-1761. *Letter.* 1744. 1 item (4 p.)

A.L.S. (1744 Mar. 4, Bologna) concerning his enemies who are working to prevent his appointment; in Italian.   MSS 954 A

201. Gould, William, 18th cent. *An account of English ants which contains their different species and mechanisms; their manner of government and a description of their several queens; the production of their eggs and process of the young; the incessant labours of the workers or common ants with many other curiosities observable in these surprising insects.* 1745. [3], 100, [2] p.; 22 cm.

Manuscript of a work published in 1747.   MSS 226 B

202. Hamberger, Georg Erhard, 1697-1755. *Letters.* 1745-1755. 2 items.

A.L.S. (1745 July 26, Jena) giving expert opinion on a fatal case of probable poisoning, and A.L.S. (1755 Jan. 5, Jena) with thanks for his admission to the Thürmainzische Academie der Nützlichen Wissenschaften; in German.   MSS 658 A

III. Drawings of the magic square and symbols from *Miscellanea alchymica et astrologica* (16th century).
Entry No. 20, SI Neg. 84-7178.

203. Bassi Verati, Laura Maria Caterina, 1711-1778. *Letter.* 1746. 1 item (1 p.)

A.L. (1746 Nov. 1) invitation to Dr. Gledini; in Italian.

MSS 50 A

204. Hollmann, Samuel Christian, 1696-1787. *Letters.* 1747. 2 items.

Two A.L.S., one (1747 Dec. 7, Göttingen) a receipt for textbooks, the other (1747 June 9, Göttingen) submitting a review for publication; in German.   MSS 713 A

205. Lewis, William, 1714-1781. *Receipt.* 1748. 1 item (1 leaf)

A.D.S. (1748 Mar. 30, London) acknowledging receipt of one guinea from Sir Hugh Smithson for the first six numbers of *The Philosophical Commerce of Arts.*

MSS 890 A

206. Mazéus, Jean Mathurin, 1713-1801. *Philosophiae pars [quarta] seu mathesis.* 1748. 345, [9] p.: ill.; 18 cm.

On geometry; 13 folded leaves of figures. Colophon dated (1748 Dec. 16) and signed: Louis [Casimirre?] Parent. Spine title: Mazeus / Phil. et Math / 1748. Proemium in Latin, text in French. Manuscript description attached.

MSS 962 B

207. Simpson, Thomas, 1710-1761. *Papers.* 1748. 2 items.

A.L.S. (1748 May 18) to Wm. Jones enclosing holograph ms. S. (1748 May 26, Woolwich): "Of the Fluents of Multinomials and Series affected by Radical-Signs, which do not begin to converge till after the 2d Term--In a Letter to Wm. Jones Esq. by T. Simpson" (5 p.). Holograph ms. annotated in a different hand with "XI" added before title and "F.R.S." after Simpson's name.

MSS 1376 A

208. Sulzer, Johann Georg, 1720-1779. *Papers.* 1748-1773. 3 items.

A.L.S. ([17]48 July 23) in German, autograph signature (1773 Sept. 9) on recto of leaf with epigram S. by Christianus Augustus Clodius (1773 May 11) in Latin, and autograph epigram in Italian signed "Giann. Sulzer."

MSS 1436 A

209. J. B. *Beredeneerd onderwijs in de wiskunde door J. B.* 1749. 169 p.: ill; 21 cm.

At foot of title page: W.W.P. Sc., 1749.   MSS 244 B

210. Koenig, Samuel, 1712-1757. *Letter.* 1749. 1 item (4 p.)

A.L.S. (1749 Oct. 10, la Haye) possibly to Haller, concerning Leibniz and Wolff; in French.   MSS 791 A

211. Montferrier, Alexandre André Victor Sarrazin de, 1729-1863. *Document.* [between 1749 and 1863]. 1 item (2 p.)

A. ms. (undated) concerning publication of a book of his with a printed notice of the book attached; in French.

MSS 1030 A

212. *Astronomy, Mechanics, and Optics.* [2nd half of 18th Century?]. 72 leaves: ill.; 46 cm.

Notebook with 22 sections on astronomy, the mechanics of motion and equilibrium, and optics, including telescopes and the rainbow. "Hamilton" written at the head of one leaf. Pen and ink diagrams. Blank leaves separate sections. Spine title: Mechanics, Optics, Etc.

MSS 910 B

213. Bossut, Charles, 1730-1814. *Papers.* [between 1750 and 1814]. 2 items.

A.L.S. and holograph note; in French.   MSS 151 A

214. Cullen, William, 1710-1790. *Of the nervous system; of the action of moving fibres; of the functions of the brain.* [ca. 1780]. 50 p.; 20 cm.

Spine title: Cull. Nerv. Syst.   MSS 210 B

215. *Des differentes especes de corps.* [2nd half of 18th century]. [91] p.; 30 cm.

Manuscript on physics including sections on fire, color, light, optics, water, the magnet, and the system of the world. A section on "electricité naturelle" cites the work of Abbé Nollet in 1749 and of Franklin; in French. Spine title: Physiques-l'Electricité--Optique etc. / (1770). One cm. strip cut from top of first page above title.

MSS 1278 B

216. *Electricity.* [1750?]-1801. 8 p.; 21 cm.

With a lecture on fractures and brain surgery, dated Feb. 28, 1801, (13 p.). Spine title: Electricity - Manuscript.

MSS 215 B

217. Folkes, Martin, 1690-1754. *Letter.* 1750. 1 item (2 p.)

A.L.S. (1750 June 29, London) to the Académie Royale des Sciences with elaborate thanks for the gift of 69 volumes of "Histoire et Memoires" of the Académie; in French.   MSS 524 A

218. *In universam phil[osoph]iam commentarii.* [2nd half of 18th century]. [301] p.: ill.; 22 cm.

A text on physics in four parts: I. De corpor[um] principiis (which reviews the work of Gassendi, Descartes, and Newton), II. De corporum proprietatib[us], III. De corporum affectionibus, and IV. De corporib[us] in particulari. Specific topics include motion and magnetism; in Latin. Spine title: Physic[a] / Elemen[ta]   MSS 1253 B

219. Knight, E. *Arithmetic.* [1750-1850]. [178] p.: ill.; 21 cm.

Spine title in later hand: Knight, Arithmatic(!) c. 1790. In the original hand on front cover: E Knight. Topics include weight, interest, square and cube roots. Illustrations include diagrams and decimal tables.   MSS 838 B

220. Knight, E. *Geometry.* [1750-1850]. [180] p.: ill.; 21 cm.

Spine title in later hand: Knight, Geometry, c. 1790. Contents include geometrical theorems and problems, trigonometry and surveying, and the use of the plain table, the Theodolite, and the circumferenter. Illustrations include diagrams and tables of logarithms, sines, and tangents.   MSS 839 B

221. Messier, Charles Joseph, 1730-1817. *Letter.* [between 1750 and 1817]. 1 item (1 leaf)

A.L.S. (Mar. 26) to Cotte concerning a letter from Rozier about navigational instruments; in French.

MSS 995 A

222. *Notes on natural philosophy.* [1750-1800?]. 2 v. (487 p.): ill.; 20 cm.

Topics include mechanics, motion, electricity, magnetism, and optics. The section on astronomy in vol. 2 is incomplete; 123 blank pages follow. Also included are tables, one laid in, and 29 folded plates of diagrams. Title page: Notes / on / natural / philosophy / in two volumes / Vol. 1 / Aberdeen. Spine title: Natural philosophy.

MSS 1062 B

223. *Specious arithmetick.* [between 1750 and 1850?]. 77 leaves: ill., some col.; 19 cm.

Bound ms. in 4 parts: "Notation," "Elements of Geometry," "Problems" and "Definitions." Spine title: Math.

MSS 1285 B

224. *Drawings of Sundials.* [between 1750 and 1850?]. 1 item (2 leaves): col. ill.

Two annotated ink and wash drawings of a sundial: "Horizontal Dial for the Latitude 50" and "Vertical Dial turned to the East for the Latitude 50." [VIII]

MSS 1437 A

225. Vecchi, Domenico de. *Memoria sopra un nuovo termometro del [--?--] Domenico Vecchie; [Memoria]; [Correzione della altezze barometriche per mezzo del termometro]; [Memoria sopra un nuovo magnetometrografo].* [between 1750 and 1850]. [58] p., 7 leaves of plates: ill.; 24 cm.

Bound ms. containing four texts. The second concerns the barometer; in Italian. First treatise numbered "XXIII," fourth numbered "XXVI."   MSS 1304 B

226. Wilson, William, fl. 1750. *[Trigonometry]; [Navigation]; [A journal by God's permission of the ship* Panther *from England toward Madeira, Capt. J[oh]n Bonline Commander] [by William Wilson].* [1750]-1753. [166] p., [50] leaves: ill., map; 25 cm.

Bound holograph ms. signed, containing three texts, one a ship's log (1753 Nov. 14-Dec. 4). Spine title: Navigation. One or more pages missing from front of codex; text begins with "Case II." Also, the page for "Case III" has been torn out. Front cover signed twice.   MSS 1310 B

227. Bernoulli, Jean, 1710-1790. *Letters.* 1751-1757. 2 items.

Two A.L.S. to Gesner concerning books purchased by Bernoulli for Gesner; in French. MSS 97 A

228. Clairaut, Alexis Claude, 1713-1765. *Letters.* 1751-1753. 2 items.

A.L.S. (1751 Feb. 4, Paris) to an unidentified correspondent concerning Clairaut's "Elemens de Geometrie et d'Algebre" in French; and A.L.S. (1753 April 6, Paris) thanking Richardson for his kindness during Clairaut's visit, in English. MSS 348 A

229. *Geometrical problems for the construction of dials; Tables and rules for finding Easter, gold[en] numbers, epact, dom[inical] letter, &c,&c,&c, vol. 2d. carefully transcribed by William Walker near Old Meldrum.* 1751. 154, 74 p.: ill. (some col.); 33 cm.

Manuscript lacks p. 1-6. Three engraved compasses on paper laid in. MSS 269 B

230. Simson, Robert, 1687-1768. *Letters.* 1751-1756. 2 items.

A.L.S. (1751 Nov. 18, Glasgow) to Joshua Sharpe concerning University finances, and A.L.S. (1756 Nov. 5, Glasgow) containing a transcription of an A.L.S. from William Rouet concerning University finances. MSS 1377 A

231. Strömer, Märtin, 1707-1770. *Letter.* 1751. 1 item (4 p.)

A.L.S. (1751 Apr. 2/13, Upsala) concerning the Académie des Sciences, his own astronomical observations, and the professional assistance of La Caille and Romé de L'Isle; in French. MSS 1427 A

232. Buffon, Georges Louis Leclerc, comte de, 1707-1788. *Letter.* 1752. 1 item (3 p.)

A.L.S. (1752 July 22) to Ruffey concerning payment of a debt, and experiments with lightning; in French. Additional material includes photocopies and translations. MSS 201 A

233. Kästner, Abraham Gotthelf, 1719-1800. *Papers.* 1752-1783. 5 items.

A.L.S. (1752 Jan. 1) in Latin, autograph signature with epigram (1783 Dec. 29) A.D.S., and 2 A. sheets of notes; in German. MSS 771 A

234. Lagrange, Joseph Louis, comte, 1736-1813. *Letters.* 1752-1812. 7 items.

Six A.L.S. to various correspondents including Arago and Maraldi and D.S.; in French. MSS 811 A

235. Rittenhouse, David, 1732-1796. *Papers.* [between 1752 and 1796] 2 items.

A.L.S. (undated) to His Excellency the President, conclusion and signature only, and D.S. (1796 May 24, Philadelphia) testifying to the qualifications of Mr. John Hall as surveyor, also signed by Ewing, Patterson, and Ellicott. MSS 1217 A

236. Wachendorff, Everardus Jakob van, 1702-1758. *Correspondence.* 1752. 1 item (3 p.)

A.L.S. (1752 Nov. 13) to Ludwig (2 p.) and Ludwig's reply on the third page; in Latin. MSS 1519 A

237. Bory, Gabriel de, 1720-1801. *Passages du soleil et de Mercure aucx fils horizontaux et verticaux du réticule de la lunette perpendiculaire du sextans, observés le 6 mai 1753.* 1753. 1 item (12 p.)

Holograph ms. in French. MSS 147 A

238. Gralath, Daniel, 1739-1809. *Document.* 1754. 1 item (4 p.)

A.D.S. to chairmen of various sections of Naturforschende Gesellshaft zu Danzig to send an outline of their proposed lectures, with their answers including Reyger and Sendel; in German. MSS 610 A

239. Jussieu, Bernard de, 1699-1777. *Review.* 1754. 1 item (4 p.)

Holograph ms. (1754 Sept. 6) reviewing Brisson's *Le règne animal divisé en neuf classes, première classes des quadrupedes,* signed by Jussieu and Duhamel de Monceau; in French. MSS 765 A

240. Gliffard, Joseph. 18th cent. *Cosmographia.* 1755. 123, [327] p.: ill.; 19 cm.

MSS 229 B

241. Lalande, Joseph Jerôme Le Français de, 1732-1807. *Papers.* 1755-1806. 20 items.

Seventeen A.L.S. to various correspondents including Arduino, Beccaria, and Deluc, 2 A.D.S., and a D.S.; in French. Additional material in second folder.

MSS 814 A

242. Algarotti, Francesco, conte, 1712-1764. *Letter.* 1756. 1 item (1 p.)

A.L.S.? (1756 June 1, Venice) to Maupertuis concerning M. de la Condamine; in French. MSS 72 A

243. Fontana, Josephus, 18th cent. *Elementa geometriae tum planae, tum solidae a Joanne Camera Plumbinensi semis sondiorum causa commorante, exaraba, abque ab admodum Re'do Do[mi]no Josepho Fontani S. & Theologiae Doctore explicata A.D. 1756.* 1756. 302, 123 p., 15 leaves of plates; ill.; 21 cm. MSS 266 B

244. Hales, Stephen, 1677-1761. *Letters.* 1756-1758. 2 items.

A.L.S. (1756 Mar. 8) concerning ventilators on naval ships and purification for bad-tasting milk, and A.L.S. (1758 Mar. 22, Teddington) to Rev. Mr. Wetstein, concerning arrangements for administering the sacrament.

MSS 654 A

245. Revelli, Philippo. 18th cent. *Arithmetices, algebrae ac geometriae institutiones.* 1757. [1], 380 p.: ill.; 24 cm.

Title page: Gasparis Antonii Manera clerici / 1757. Marginal computations and diagrams; in Latin.

MSS 948 B

246. Speri, Joseph, 18th cent. *Elementa phisicae ab admodum r[evere]ndo do[mi]no Joseph Speri in Seminario Archiepiscopali S. Georgii philosophiae et S. theologiae lectore, preclarissime explicata, ac a Joanne Camera Plumbinensi exarata.* 1757. 439 [i.e., 436] p.: col. ill.; 21 cm.

Bound ms. containing "Index Capitum" at end and folded plate of diagrams in black and red ink. Topics include electricity and the magnet; in Latin. Spine title: Phisica. MSS 1286 B

247. Académie des sciences (France). *Attendance list.* 1758. 1 item (3 p.)

Attendance list (1758 Jan. 25) containing signatures of members attending the meeting on that date, including Jussieu, d'Alembert, and Buffon; in French. MSS 17 A

248. Gaubius, Hieronymus David, 1705?-1780. *Letter.* 1758. 1 item (1 p.)

A.L.S. (1758 Dec. 1) to Allamand concerning a balsam of remarkable properties; in French. MSS 572 A

249. *Navigazione.* 1758. [49] leaves: ill.; 35 cm.

With short philosophical treatise on God: Oppinioni filosofiche d'Iddio; in Italian. Treatise on theoretical and practical navigation, with diagrams and navigational charts. Spine title: Navigazione / 1758. MSS 970 B

250. Oberkampf, Christoph Philipp, 1738-1815. *Essais faits à la manufacture de Joui près Versailles machine de Charpentier pour craticuler et [perier?] des plaques de cuivre par Mr. Oberkampf.* 1758-1815. 1 item (2 p.): ill.

MSS 1082 A

251. Barker, Joseph, 18th cent. *[Navigation].* 1759-1761. [197] p.: ill., maps; 33 cm.

MSS 285 B

252. Cousin, Jacques Antoine Joseph, 1739-1800. *Letter.* [between 1759 and 1799]. 1 item (1 p.)

A.L.S. to Saint-Léger concerning an article by Condorcet; in French. MSS 378 A

253. La Harpe, Jean-François de, 1739-1803. *Letter.* [between 1759 and 1803]. 1 item (2 p.)

A.L. (April 15); in French. MSS 812 A

254. Beccaria, Giovanni Battista, 1716-1781. *Letter.* 1760. 1 item (1 folded leaf)

A.L.S. (1760 Oct. 25, Turin) to Joseph Maria Beccaria, Mondovi; in Italian. MSS 54 A

255. Calandrelli, Giuseppe, 1749-1827. *Correspondence.* 1760-1818. 45 items: ports.; 34 cm.

Twenty-two A.L.S. from Calandrelli to various correspondents (1778-1818) bound in one volume; 22 A.L.S. from Bonati to Calandrelli (1786-1791) bound in one volume; and 1 A.L.S. from Bonati to an unidentified correspondent (1760 May 7); all boxed together; in Italian. On box: Teod. Bonati Gius. Calandrelli. German typescript summaries in box. MSS 282 B

256. Forcheron, D., 18th cent. *[Physica].* 1760. 315 p.: ill.; 26 cm. MSS 247 B

257. Gessner, Johann, 1709-1790. *Papers.* 1760-1783. 2 items.

A.L.S. (1760 Mar. 22, Zurich) to Ludwig, in German, and autograph signature (1783 June 14); in Latin. MSS 591 A

258. *Trattato dell'arithmetica superiore; [Trattato della trigonometria]; [Trattato della meccanica].* [ca. 1780]. 304 p.: ill.; 22 cm.

Bound ms. containing three treatises, including 5 folded plates of pen and ink drawings. The first and second treatises are in 3 parts each, and the third treatise is in 7 parts; in Italian. Box spine title: Mathematica / Meccanica. MSS 1300 B

259. Alembert, Jean Lerond d', 1717-1783. *Papers.* 1761-1781. 12 items.

Ten A.L.S. (including one to Court de Gebelin); 1 dictated letter; and a ms. copy of corrections and additions for the new ed. of Elémens de musique with additions and remarks in Alembert's handwriting. MSS 71 A

260. Jacquier, Françios, 1711-1788. *Papers.* 1761. 2 items.

A.L.S. (1761 May 25, Rome) to Algarotti in French, and undated A. ms. in Italian. MSS 746 A

261. Lichtenberg, Georg Christoph, 1742-1799. *Letter.* [between 1762 and 1799]. 1 item (4 p.)

A.L.S. (undated); in German. MSS 893 A

262. Lambert, Johann Heinrich, 1728-1777. *Letters.* 1763-1772. 2 items.

A.L.S. (1763 July 19) to Gesner concerning his travels to Chur and Valtellina, and A.L.S. (1772 Jan. 29); in German. MSS 817 A

263. Lyons, Israel, 1739-1775. *Letter.* 1763. 1 item (1 leaf)

A.L.S. (1763 Oct. 11, Cambridge) to Mr. da Costa presenting 2 books for the Royal Society Library. MSS 934 A

264. Colle, Francesco Maria, 1744-1815. *Poem.* [between 1764 and 1815]. 1 item (4 p.): col. ill.

Holograph poem S. to the Duchess of Chartres with drawing of double-acting pump, wing rack and pinion drive; in French. MSS 360 A

265. Linné, Carl von, 1707-1778. *Letter.* 1764. 1 item (2 p.)

A.L.S. (1764 Dec. 23, Upsalia) to Duchesne; in Latin. MSS 904 A

266. Le Monnier, Pierre Charles, 1715-1799. *Papers.* 1765-1809. 3 items.

D.S. (1781 May 21, Paris) receipt for 100 copies of his paper on the eclipse of 1778, a portion of a letter in a fair hand (probably 1765 July 3) probably to Matteucci concerning the transit of Venus, 1761, and 17 sheets of daily barometric and thermometric observations (1791-1809), not in Le Monnier's hand; in French. MSS 877 A

267. *Principes sur la sphere où l'on traitera d'abord du systhême de Ptolomée et ensuite de celui de Copernic avec quelques notions abregés sur l'astronomie, année 1765.* 1765. 152 p.; 18 cm.

Bound ms. in French. MSS 1259 B

268. Montucla, Jean Etienne, 1725-1799. *Letter.* 1766. 1 item (4 p.)

A.L.S. (1766 Nov. 26, Paris) to an unnamed "madame" concerning a property dispute; in French.   MSS 1036 A

269. Vaucanson, Jacques de, 1709-1782. *Papers.* 1766-1777. 5 items.

Three A.L.S., two to Deidier and one to "Messieurs les freres jubié," ms. (8 p.): "De la filature des soies. . . par M. Vaucanson," and ms. (published in 1770): "Explication des figures du tour(?) la sole par M. Vaucanson," with 9 engraved plates (Plates I-VI and soiled duplicates of three of these) dated 1770; in French.   MSS 1503 A

270. Boškovič, Rudjer Josip, 1711-1787. *Document.* 1767. 1 item (1 p.)

Signed and annotated document (1767 April 2, Pavia); in Italian.   MSS 150 A

271. Titius, Johann Daniel, 1729-1796. *Letter.* 1767. 1 item (2 p.)

A.L.S. (1767 Oct. 6, Wittenberg) to a professor; in German.   MSS 1472 A

272. Orloff, Wladimir, 1742-1812. *Letter.* 1768. 1 item (1 leaf)

A.L.S. (1768 June 13, St. Petersburg) to Haller; in German.   MSS 1089 A

273. Sanxai, Glicerius, 18th cent. *Dissertatio de re electrica tradita anno 1768 lectore P. Glicerio Sanxai Ordinis Scholarum Piarum.* 1768. ccciii p.; 22 cm.

Manuscript, possibly holograph, in Latin with extensive quotations in Italian. Bookplate of conte Paolo Vimercati-Lozzi.   MSS 1274 B

274. Waddington, R. E., 18th cent. *Letter.* 1768. 1 item (3 p.)

A.L.S. (1768 July 8) to Piggott concerning shipping a telescope and magnetism.   MSS 1520 A

275. Boeckmann, Johann Lorenz, 1741-1802. *Papers.* 1769-1795. 10 items.

Nine A.L.S. to various correspondents including Beckmann and holograph outline of Boeckmann's book on mechanics published at Karlsruhe, 1769; in German.   MSS 127 A

276. Canterzani, Sebastiano, 1734-1810. *Letters.* 1769-1787. 3 items.

Three A.L.S. to Calandrelli, Ferrari, and an unidentified correspondent; in Italian.   MSS 298 A

277. Gerstenbergck, Johann Laurentius Julius von, 1749-1813. *Letter.* [between 1769 and 1813]. 1 item (1 p.)

A.L.S. concerning a review of his new work in the *Intelligenzblatt* and other works ready for printing; in German.   MSS 589 A

278. Gilpin, Thomas, 18th cent. *Drafts of surveys, sketches, and calculations of improvements by Thomas Gilpin, gent.* 1769-1770. 17 items (bound together): ill., col. folded maps; 54 cm.

Thirteen ms. sheets, illustrated with topographic sketches; 3 colored and folded ms. maps; and 1 printed broadside; concerning surveys in Pennsylvania and Maryland for canal improvement.   MSS 290 B

279. Smith, Mary, 18th cent. *[Commonplace book concerning science and mathematics]* [between 1769 and 1780?]. [303] p.: ill. (one col.); 33 cm.

A. ms. containing an index to works by Robinson, Rowning, and Jones and Baxter's *Matho*, notes, abstracts, transcriptions from journals (1737-1769) with Smith's critiques. Topics include the Copernican System, Newton's works, astronomy, the telescope, small pox, the barometer, Hebrew terms, and the work of Jones. Three pages of notes laid in. Bookplate: Mary Smith, Thorney Abbey.   MSS 1281 B

280. Bussius, Petrus, 18th cent. *Ad phisica introducio.* [1774?]. 304 p.: folded ill.; 22 cm.   MSS 243 B

281. *[Cahiers de l'arithmétique].* [ca. 1780-90]. 229 p.; 24 cm.

Spine title. MSS 267 B

282. Englefield, Henry Charles, Sir, 1752-1822. *Drawing.* 1770. 1 item (1 p.); ill.

Drawing of the geometrical construction of the eclipse of the moon, Oct. 11, 1772. [IX] MSS 484 A

283. *[Farragine matematica].* [between 1770 and 1781]. [72] p.; 27 cm.

Spine title: Math. 1770-81. In red and black ink. MSS 261 B

284. Lous, Christian Carl, 1724-1804. *Arithmetica, Algebra, Geometrien, transformatio, trigonometrien, landmaalingen.* [177-?]. [400] p. ill. (some col.), maps; 36 cm.

Each section begins with heading, "Den Förste grund af. . ." profusely illustrated in ink and watercolor, including a folded diagram of a boat and a folded map. Bound, stamped in gilt: title on spine is "Mathematics," "C.S. Lous" is stamped on front cover, "1822" on back cover. "C.S. Lous" is also prominent in the original handwriting in the diagram of a boat; in Danish. MSS 830 B

285. *Nachricht von einer neü erfünden aequinoctial sonnen-uhr.* [ca. 1780]. 1 item (22 p.)

A. ms. announcing a new equinoctial sundial for sale; in German. Dealer's description laid in. MSS 1064 A

286. *[Physics -- experimental equipment].* [between 1770 and 1830?]. 1 item (24 leaves): all col. ill.

Water colors of experimental equipment used in physics, e.g., pulleys, magnets, weights, fountains, and electricity. On paper watermarked Sebille, van Ketel, & Wassenbergh. "A. Volta" is written on one leaf, but is believed not to be Volta's autograph, though contemporary. [X] MSS 1136 A

287. Saussure, Horace Bénédict de, 1740-1799. *Physica excerpta ex publicis praelectionibus D.D. Hor. Ben. De Saussure in Genevensi Academia Professoris celeberrimi.* 1770-[1776?]. 1-275, 293-370 [i.e., 351] leaves: ill.; 24 cm.

Manuscript on physics with topics including the elements, hydrostatics, magnetism, and gravity; in Latin. Spine title: physica t[omus] I Generalis / Prolectiones de Hon.(!) Ben. De Saussure. MSS 1275 B

288. *Tractatus in quo de vero partiu[m] universi situ et motu seu de vero mundi systemate inquiritur; [De astronomia positiva]; [Tractatus de lumine]; [Tractatus de cosmographia]; [Proemium in phisicam].* 1770. [515] p.: ill., map; 21 cm.

Bound ms. with 5 texts, the 4th with a colophon: "Finis est. Proficiat hac 21 augusti 1770 in celeberimo Castrensi Pedagogio a doctissimo D.D. Verheyden Professore Secundario." 14 plates signed "C. H. Becker" (2 dated 1766) and 1 plate signed "Apud P. Denique Lovanii;" in Latin. Spine title: Pars Physices. First page stamped "Rev. T. Flynn." MSS 1296 B

289. Weigleb, Johann Christian, 1732-1800. *Letters.* 1770-1772. 3 items.

Three A.L.S., one to Monsieur Nicolai; in German. MSS 1565 A

290. William, Thomas. *[Measuring the earth].* [between 1770 and 1788]. 16 p.: ill.; 22 cm.

Holograph ms. S. with two-part text: "Explanation of the Figure, Page 7th," and "Explanation of the Figure, Page 15th." Williams disagrees with Sir Isaac Newton and "the late Maj. Gen. Ray" (p. 3; cf. p. 16: Maj. Gen. Reg.(?)) on how to measure the earth's diameter. Possibly preliminary notes for his *Method to discover the true diameters of the earth.* MSS 1308 B

291. Bézout, Etienne, 1730-1783. *Letters.* 1771-1783. 2 items.

Two A.L.S. (1771 July 13, Bayonne; 1783 Aug. 14, Paris); in French. MSS 111 A

292. Priestley, Joseph, 1733-1804. *Correspondence.* 1771-1801. 6 items.

Five A.L.S. (1771-1801) to various correspondents including Carey, Lindsey, Turner, and Wedgwood, 1 A. ms. (undated) in French, a draft of a complimentary letter from the students of chemistry, medicine, and pharmacy in Paris, beginning "Les chemistes de Paris au Docteur Priestley, Salut," perhaps with autograph revisions by Lavoisier; in English.　　　　MSS 1174 A

293. Baumé, Antoine, 1728-1804. *Letters.* 1774-[1801]. 2 items.

A.L.S. (1774 March 17) billing Perregaux for medicines and A.L.S. (an 9 fructidor 16, i.e., 1801 Sept. 3); in French.　　　　MSS 53 A

294. Cassini, Jean Dominique, comte de, 1748-1845. *Papers.* 1772-1817. 5 items.

Two A.L.S., one to Messier; A.D.S. recommending an armillary sphere devised by Robert de Vaugondy; A.D.S. for receipt of money from the Legion d'honneur treasury; and a ms. document, signed Louis Antoine, duc d'Angoulême, recommending the reestablishment of Cassini's pension; in French.　　　　MSS 312 A

295. *Tractatus de corpore naturale; [Tractatus de motu]; [Tractatus de mechanica].* 1772. 123 [i.e., 246]. p.: ill.; 21 cm.

Bound ms. containing three treatises with separate colophons, the 1st dated (1772), and 7 engraved plates of figures signed C. H. Becker ex lov. Spine title: Corpus, motus et mechanic[a].　　　　MSS 1294 B

296. Belgrado, Jacopo, 1704-1789. *Letter.* 1773. 1 item (1 p.)

A.L.S. (1773 June 11, Modena) acknowledging receipt of a letter from Alberti and thanking him for his kindness; in Italian.　　　　MSS 85 A

297. Born, Ignaz, Edler von, 1742-1791. *Papers.* 1773-1780. 7 items.

Six A.L.S. to various correspondents including Hermann, and a holograph ms. signed (1780 Nov. 7) reporting to the Austrian Emperor on the Imperial Montanistic Works and the Mint in Biebenbuergen; in German.　　　　MSS 143 A

298. Condorcet, Jean-Antoine-Nicholas de Caritat, marquis de, 1743-1794. *Letters.* [1773?]-1789. 2 items.

Two A.L.S. ([1773?], 1789 Feb. 4), one addressed to Cotte, concerning books; in French.　　　　MSS 365 A

299. Maskelyne, Nevil, 1732-1811. *Letters.* 1773-1806. 7 items.

Six A.L.S. to various correspondents including Buchan and Pigott, and 1 A.L.　　　　MSS 972 A

300. Messano, Augustino A., 18th cent. *Physica experimentalis.* 1773-1774. [340] p.: ill.; 25 cm.

Part 3 of Messano's *Philosophia Eclectica.* Title taken from p. [278]. Scribe identifies himself on title page as Petrus ab Albavilla and gives date (1774). First page of text dated (1773 Sept. 8). Fifteen plates of figures signed by Lucchesini; in Latin.　　　　MSS 966 B

301. Stewart, Dugald, 1753-1828. *Letter.* [between 1773 and 1828]. 1 item (1 leaf)

A.L.S. (July 31, Caimincie) to an unnamed correspondent concerning a packet for Mr. Alexander Lawrie, a bookseller in Edinburgh.　　　　MSS 1421 A

302. Coggeshall, Henry, 1623-1690. *Coggeshal's sliding rule: the description and use of the sliding rule.* 1774-1776. 55 p.: ill.; 20 cm.

On cover: Vol. 3., Nov'r. 25, 1774, R.T. On spine: R.T., Coggeshal's Sliding rule, 1774-6.　　　　MSS 214 B

303. Haller, Albrecht von, 1708-1777. *Letter.* 1774. 1 item (1 p.)

A.L.S. (1774 Oct. 15, Bern) to Caldani, Padua; in Latin.　　　　MSS 656 A

304. Karsten, Wenceslaus, Johann Gustav, 1732-1787. *Letter.* 1774. 1 item (1 p.)

A.L.S. (1774 Mar. 17, Luetzow); in German.　　　　MSS 770 A

305. Kok, H., 18th cent. *Physica.* 1774. 2 v. (488 p., variously numbered): ill.; 19 cm.

Vol. one is bound, has separately numbered parts on, e.g., motion, gravity, air, and sound; most sections conclude with folded engravings signed: C. H. Becker: Ex Lov.; and illustrations. The date 1774 appears on three title pages. Vol. two, titled "Quaestiones Calefactoris" (i.e., topics for the eager), is unbound with separately numbered sections on motion, mechanics, gravity,

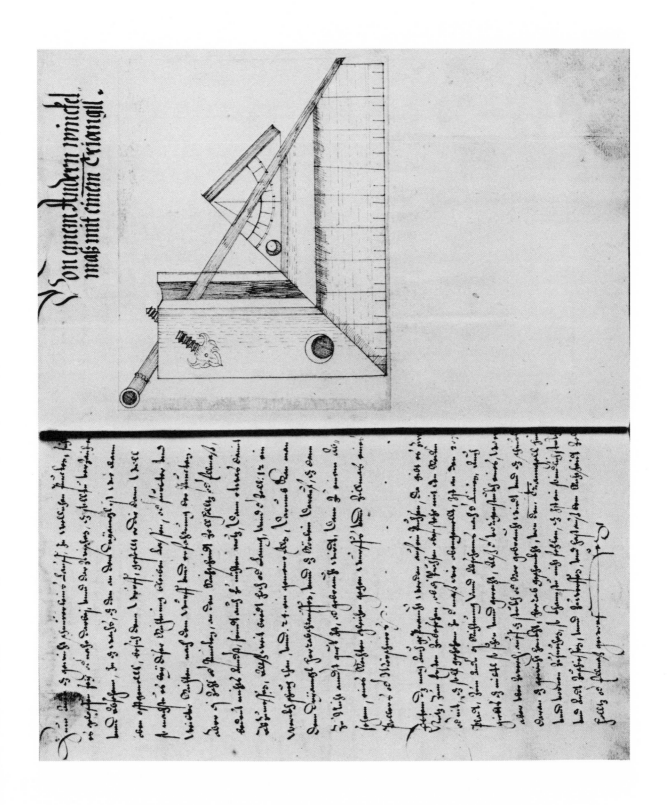

IV. Illustration from a German treatise on master-gunnery (1589).
Entry No. 33, SI Neg. 82-4831.

spheres, hydrostatics, etc. Leaves are various sizes, folded to 19 x 12 cm.; in Latin. MSS 834 B

306. Raspe, Rudolf Erich, 1737-1794. *Memorial d'un mondain [extract].* [between 1774 and 1794]. 1 item (2 p.)

A. ms. S. (undated), an extract from *Memorial d'un mondain,* Vol. 1, p. 7, with a Latin inscription (1733 Oct. 15) and an English translation of the Latin. MSS 1192 A

307. Sheridan, Elizabeth Ann Linley, 1754-1792. *Poem.* [between 1774 and 1792]. 1 item (2 p.)

A. poem "By Mrs. Sheridan on her brother's violin." MSS 903 A

308. *A Brief history of electricity.* [last quarter of 18th century]. [136] p., [4] folded leaves of ill.; 21 cm.

Folded leaves of ill. numbered 3-6. Spine title: Electricity. MSS 212 B

309. Hell, Maximilian, 1720-1792. *Letter.* 1775. 1 item (2 p.)

A.L.S. (1775 Aug. 1) recommending a person applying for an occupation at the Danish court; in German. MSS 681 A

310. Moreni, Marianno, 18th cent. *Fisica gener[a]le e sperim[enta]le lecta a P. Maria[n]no Moreni.* 1775-1776. 300, [5] p.; 22 cm.

Chiefly concerns fire and electricity; in Italian. Latin title added on title page: Physica generalis et experimentalis. Title page: Anno 1775-1776. MSS 963 B

311. Priestley, Joseph, 1733-1804. *Dr. Joseph Priestley: a collection of autograph letters, portraits, etc.* 1775-1878. 76 leaves: ill., port.; 34 cm.

Eleven A.L.S. (1775-1796) (3 to Banks), 3 fragments of holograph S. (1775-1796) (3 to Banks), 3 fragments of holograph S. (1795) of autobiography, 9 A.L.S. (1844-1878) mostly to Davis, all concerning Priestley from various correspondants including Peale, several engraved and photograph portraits, and material pertaining to the 1874 Priestley Centen, collected in bound scrapbook. MSS 1257 B

312. Spielmann, Jacob Reinbold, 1722-1783. *Letters.* 1775-1781. 2 items.

Two A.L.S. (1775 Apr. 25 and 1781 Oct. [9?]) to M. Jaquin, Conseille des Mines; in Latin. MSS 1399 A

313. D'Arquier de Pellepoix, Augustin, 1718-1802. *Letter.* 1776. 1 item (3 p.)

A.L.S. (1776 June 28, Toulouse) to Fouché, the secretary of the Académie royalle des science, describing a new and beautiful comet, and mentioning Lalande; in French. MSS 403 A

314. Ferroni, Pietro, 1744-1825. *Papers.* 1776-1822. 16 items.

Seven A.L.S. (1776-1822), 4 A. papers including a book list, and 5 letters to Ferroni from various correspondents; in Italian. MSS 507 A

315. Magalhães, João Jacinto de, 1722-1790. *Letters.* 1776-1777. 2 items.

A.L.S. (1776 Sept. 12) to John Walsh offering to be of assistance in Holland, and A.L.S. (1777 April 22, London) to M. Duchaine(?) concerning some grain, both signed "Magellan"; in French. MSS 945 A

316. Segner, Johann Andreas von, 1704-1777. *Autograph signature.* 1776. 1 item (1 leaf)

Autograph signature (1776 Mar. 28) following Latin epigram. MSS 1351 A

317. Semler, Johann Salomo, 1725-1791. *Papers.* 1776-1789. 2 items.

Autograph signature (1776 May 9, [Halle?]) containing Latin epigram, and A.L.S. (1789 Oct. 8) containing Greek quotations from the *Iliad* attached to an envelope and a Latin list; in German. MSS 1354 A

318. Achard, Franz Karl, 1753-1821. *Letters.* 1777-1779. 2 items.

A.L.S. (1777 Dec. 14, Berlin) from Achard to Christian Sendel, in German. A.L.S. (1779 Aug. 28, Berlin) from Achard to unidentified "Monsieur," in French. Supplementary ms. and typescript enclosed. MSS 9 A

319. Beckmann, Johann, 1739-1811. *Addenda.* [between 1777 and 1811]. 1 item (4 p.)

A. ms. of additional remarks to Beckmann's *Beiträge zur Ökonomie, Technologie, Polizei, und Kameralwiss* (1777-1790); in German. MSS 57 A

320. Klügel, Georg Simon, 1739-1812. *Papers.* 1777-1794. 2 items: ill.

A.L.S. (1777 Aug. 25, Hemstedt) dealing with necessary adjustments of drawings in a book on optics, and autograph signature with epigram (1794 Sept. 15, Halle); in German. MSS 788 A

321. Sage, Balthazar Georges, 1740-1824. *Letters.* 1777-[1805]. 2 items.

One A.L.S. (Nivoise 8, i.e., Dec. 28, 29, or 30) to Monsieur [Raurio?] concerning a sculpture, and one A.L.S. (1777 Feb. 3): in French. MSS 1317 A

322. Bergman, Torbern Olof, 1735-1784. *Letter.* 1778. 1 item (3 p.)

A.L.S. (1778 Sept. 8, Upsala) to Macquer concerning experiments with acids and minerals; in French. MSS 92 A

323. Gall, Franz Josef, 1758-1828. *Paper.* [between 1778 and 1828]. 1 item (3 p.)

Holograph leaf (2 p.) in French, attached to a page of biography (in another hand); in German. MSS 563 A

324. Lamberg, Maximilian Joseph, Graf von, 1729-1792. *Letters.* 1778-1785. 2 items.

A.L.S. (1778 Mar. 28, Bayern) concerning Voltaire and Count Falkenstein, and A.L.S. (1785 April 4, Bruenn) concerning marble and phosphorescent stone; in German. MSS 816 A

325. Crell, Lorenz Florenz Friedrich von, 1744-1816. *Letters.* 1779-1781. 2 items.

Two A.L.S. (1779 and 1781 Jan. 27), the latter to Nicolai urging the forwarding of books and promising new mss.; in German. MSS 379 A

326. Hellwig, Johann Christoph Ludwig, 1743-1831. *Letter.* 1779. 1 item (1 p.)

A.L.S. (1779 Aug. 20, Braunschwig); in German. MSS 682 A

327. Mayer, Johann Tobias, 1752-1830. *Letter.* 1779. 1 item (4 p.)

A.L.S. (1779, Göttingen); in German. MSS 982 A

328. Nourse, John, 18th cent. *License to publish the Nautical almanac and astronomical ephemerides.* 1779. 1 item (2 p.)

Document signed and sealed by the commission appointed for the discovery of the Longitude at Sea, including Maskelyne, authorizing Nourse to publish the *Nautical Almanac*, 1781-1786. MSS 1080 A

329. Toaldo, Giuseppe, 1719-1797. *Letters.* 1779-1782. 2 items.

A.L.S. (1779 Nov. 11, Padua) to Condorcet, secretary of the Académie des sciences at Paris, in Italian, and A.L.S. (1782 Feb. 3, Padua) to Deluc praising his inventions, in French. MSS 1473 A

330. Beireis, Gottfried Christoph, 1730-1809. *Letter.* 1780. 1 item (1 folded leaf)

A.L.S. (1780 Nov. 6, Helmstädt) expressing thanks for the salary received for his lectures on natural science at the University of Helmstädt; in German. MSS 84 A

331. Faujas de Saint-Fond, Barthélemy, 1741-1819. *Letters.* 1780-1801. 3 items.

Three A.L.S., to M. Delisle on the subject of Malesherbes, to Geoffroy Saint-Hilaire, and to M. Legrand and M. Molinos; in French. MSS 502 A

332. Franklin, Benjamin, 1706-1790. *Letters.* 1780-1786. 2 items.

A.L.S. (1780 May 30, Passy) to M. Grand requesting payment to Mr. James Woodmason for paper, and A.L.S. (1786 Oct. 7, Philadelphia) to Rittenhouse requesting payment to Mr. Silas Williams for attendance in Council. MSS 539 A

333. *Geometria pratica.* 1780. 142 p.: ill. (some col.); 28 cm. Cover title and date. Copy lacks p. 55-6, 61-2, 77-8, 111-114. MSS 216 B

334. Smeaton, John, 1724-1792. *Papers.* 1780-1783. 2 items.

A.L.S. (1783 Dec. 5, Sandhoe) to John Davidson, Clerk of the Peace at Newcastle, concerning "Mr. Mylne's Report on Hexham Bridge," and a holograph note S. and copied (1780 Sept. 2) by Smeaton on "the Patent Machine invented by Mr. James Watt and Co. of Birmingham." MSS 1379 A

335. *Spiegazione della meridiana fatta nel giugno del 1780 in casa del Ecc[ellentissi]mo Sig[no]re Duca Mattei e di tutte lecose in essa contenute.* 1780. 22 [i.e., 44] p.: col. ill.; 25 cm.

Bound ms. containing a dedication in 14 Latin Hexameters entitled "De Libello hoc ad D. Philippum Matthejum" and a 55 x 41 cm. folded plate entitled "Orologii ac Meridiana Matthejana Figura subdecupla"; partly illegible note on author and hand on last page; in Italian. "Phillipps MS 7523" on dedication page. MSS 1287 B

336. Bontadosi, Girolamo, d. 1840. *Dissertazione sopra l'elasticità dell'aria letta nell'Accademia di Fisica Sperimentale di S.E. il Sig'le Cardinale de Zelada il di XXIX Aprile MDCCLXXXI da Girolamo Bontadosi.* 1781. [26] p.; 29 cm. MSS 283 B

337. Galvani, Luigi, 1737-1798. *Papers.* 1781-1797. 3 items.

Two medical certificates (1787 Sept. 23 and 1797 June 29) and a holograph leaf beginning "La Forza nervea muscolare," with statement of authentication; in Italian. MSS 568 A

338. Jeaurat, Edme Sébastien, 1725-1803. *Letter.* 1781. 1 item (1 p.)

A.L.S. (1781 Oct. 20, Paris) to the Huttons concerning Jeaurat's projected visit in four years; in French. MSS 752 A

339. Lalande, Joseph Jérôme Le Françias de, 1732-1807. *Physica specialis.* 1781. 275 [i.e., 276] p.: ill.: 20 cm.

Holograph S. (1781 Jun. 30 - see p. 275 [i.e., 276]). Topics include cosmography, physical astronomy, optics, mechanics, the geological elements, sound, and light. The second part of the section "De electricitate" (pp. 208-223) is quoted from a French source. Paragraphs are numbered: 1289-1928. Illustrations include ink drawings in black and brown and two engravings, one folded, both attributed "a Paris chez Crepy" and the first with: H. von Loon. Index on unnumbered pages at the end; in Latin. MSS 832 B

340. Playfair, Robert, 18th cent. *Letter.* 1781. 1 item (2 p.)

A.L.S. (1781 Apr. 3, Edinburgh) to Chalmer about a court decree against the Brownings in favor of Captain Napier. MSS 1149 A

341. Bernoulli, Jean, 1744-1807. *Letters.* 1782-1799. 4 items.

Four A.L.S., one to Junius; in German and French. MSS 98 A

342. Girard, Stephen, 1750-1831. *Receipt.* 1782. 1 item (1 p.)

A.D.S. (1782 Dec. 14, Philadelphia), receipt for a bag of coffee. MSS 596 A

343. Lakanal, Joseph, 1762-1845. *Papers.* [between 1782 and 1845]. 2 items.

A.L.S. (undated) to Morand, and D.S. (an 3 nivose 3, i.e. 1794 Dec. 23), also signed by Deleyre; in French. MSS 813 A

344. Rennie, John, 1761-1821. *[Abstracts and transcriptions].* 1782-1783. [229] p.: ill.; 27 cm.

Holograph ms. of abstracts and transcriptions of books and articles published in 1768-1783 by Lewis, John Helins, Nevil Maskelyne, Dr. Dobson, Dr. James Lind, M. l'Abbé Bossut, Smeaton, and M. Perronet. Three of 9 folded plates are labeled in detail in French. Topics include a water gin, evaporation, a wind gage, bridges, and mechanical power. Cover: John Rennie Memoirs 1782.

Fly leaf: written by John Rennie when a student at Edinburgh 1782.   MSS 1266 B

345. *Sopra un nuovo barometro (Lez[ion]e incompleta).* [between 1782 and 1800]. 1 item (74 p.)

Manuscript concerning the barometer written in 82 numbered paragraphs in four chapters; the heading for the Fifth Chapter is followed by 4 blank pages. The latest date cited in a footnote is 1782; in Italian. The title page begins: "#XXII." The first page of text has been marked: "No. 1."   MSS 1391 A

346. Baier, Ferdinand Jacob, 1707-1788. *Letter.* 1783. 1 item (1 folded leaf)

A.L.S. (1783 May 7, Ansbach) from Baier to Delius; in German.   MSS 36 A

347. Banks, Joseph, Sir, 1743-1820. *Papers.* 1783-1817. 21 items.

Fifteen A.L.S. including one to Harrison (March 6, London) and one to Garthshore (1783 June 22, London) concerning Pulteney's book on Linné; 2 D.S.; holograph on the method of conducting the experiment of freezing quicksilver; and 3 p. of copies of Latin and Greek inscriptions in the hand of W. H. Smyth, Capt., R.N.
                                    MSS 44 A

348. Brinkley, John, 1763-1835. *Letter.* [between 1783 and 1835]. 1 item (2 p.)

A.L.S. concerning his activities during the previous week. Sheet torn in half.   MSS 180 A

349. Deluc, Jean André, 1727-1817. *Papers.* 1783-1811. 4 items.

Three A.L.S., one to Marcet, in French, and a holograph article concerning Watt's specifications; in English.   MSS 425 A

350. Hassenfratz, Jean Henri, 1755-1827. *Papers.* 1783-1804. 5 items.

Four A. reports, one (an 12 prairial 8, i.e. 1804 May 28) "Observations sur la cause qui augmente l'intensité du son dans les porter-voix," signed by Cuvier, and A.L.S. (1783 Dec. 12, Wolfsberg) to M. Pacquet(?); in French.
                                    MSS 670 A

351. Marat, Jean Paul, 1743-1793. *Letter.* 1783. 1 item (3 p.)

A.L.S. (1783 Aug. 24, Paris) to M. de St. Laurent concerning optics; in French.   MSS 956 A

352. Montgolfier, Jacques-Etienne, 1745-1799. *Letters.* 1783-1790. 4 items.

Three A.L.S., one to Beaumarchais, and 1 A.L.; in French.   MSS 1033 A

353. Perronet, Jean Rodolphe, 1708-1794. *Letter.* 1783. 1 item (1 leaf)

A.L.S. (1783 Dec. 17, Paris) concerning his passport; in French.   MSS 1118 A

354. Gouy, Louis-Marthe de, 1753-1794. *Document.* 1784. 1 item (2 p.)

D.S. (1784 Mar. 1, Paris) agreeing to certain conditions under which Mesmer would teach the principles of animal magnetism; in German.   MSS 606 A

355. Melanderhjelm, Daniel, 1726-1810. *Letter.* 1784. 1 item (1 leaf)

A.L.S. (1784 Oct. 15, Upsala) concerning the purchase of Eimmart's collections; in French.   MSS 988 A

356. Swinden, Jan Hendrik van, 1746-1823. *Papers.* 1784-1816. 3 items.

A.L.S. (1816 April 23, Amsterdam), and D.S. (April 7) in Dutch, and A.L.S. (1784 Nov. 15, The Hague) to J. A. Deluc; in French.   MSS 1440 A

357. Volta, Alessandro, 1745-1827. *Papers.* 1784-1823. 19 items.

Two holograph mss. "Il mio [requiem?] Stramo all'anno [18]13" (a poem) and "Dall'Aria acido-marina [deflogisticata?]," 8 A.L.S. (1787-1823) to various correspondents including Araldi, Giuseppe Galletti, Gilbert (in French), Martignoni, Marzari, and his wife, one letter to Volta (1811 Dec. 31) and 8 D.S. (1784-1818, one in Latin) concerning the Università di Pavia; in Italian.
                                    MSS 1514 A

358. Watt, James, 1736-1819. *Papers.* 1784-1818. 11 items in 2 folders: ill.

Three A.L.S. (1784 Mar. 5, Sept. 13, and Sept. 9) to Deluc, one concerning computations for wood burnt in a steam engine, two A.L.S. (1791 June 9 and 1793 March 13) to Messrs. Fermin & Tostet concerning their building 2 engines and a boiler, one A.L.S. (1788 Feb. 2) to Rennie concerning a connecting rod, four A.L.S. (1792-1818) on various topics, and an anonymous: "Answers by Mr. Watt to the Objections made to his Specifications" (20 p.). Additional material in second folder. MSS 1537 A

359. Bailly, Jean Sylvain, 1736-1793. *Papers.* 1785-1791. 6 items.

5 A.L.S. including one (1790 July 8) to Necker, Minister of Finance; and a ms. treatise (9 p.) "de Mercure"; in French. Dealer descriptions and translations in folder. MSS 37 A

360. Daubenton, Louis Jean Marie, 1716-1799. *Letter.* 1785. 1 item (1 p.)

A.L.S. (1785 Nov. 20, Paris) to M. Sabarot de l'averniere concerning a wild cat; in French. MSS 407 A

361. Gervier, R. P., 18th cent. *Cours de magnetisme commencé le 4 juin 1785 le R. P. Gervier.* 1785. 3 pts. in l v. (21 leaves, 26, 2, 45 p.): ill.; 20 cm.

Index in middle of volume. MSS 222 B

362. Griswold, Matthew, 1714-1799. *Document.* 1785. 1 item (1 p.)

D.S. (1785 Aug. 5, Norwich, Conn.) to the sheriff of the county of New London, requesting that a summons be served to John Morse. MSS 625 A

363. Herschel, William, Sir, 1738-1822. *Papers.* 1785-1814. 9 items: ill.

Eight A.L.S. to various correspondents including Banks, Bonaparte, Cavallo, Deluc, and Schröter, and an autograph signature with a picture of the telescope at Slough. MSS 696 A

364. Landriani, Marsilio, conte, ca. 1751-ca. 1816. *Letters.* 1785-1788. 2 items.

A.L.S. (1785 Sept. 10) accepting an order on Electric Conductors for a lightning arrester, A.L.S. (1788 July 23, Mattlock) to Dr. Luigi Lambertenghi concerning a scientific excursion to England; in Italian. MSS 819 A

365. Mons, Jean Baptiste van, 1765-1842. *Letters.* [between 1785 and 1842]. 2 items.

A.L.S. (1842 April 26) to the Royal Management of Agriculture concerning potatoes, and A.L.S. (undated) to Döbereiner concerning the properties of phosphorus; in French. MSS 1027 A

366. Parkes, Samuel, 1759-1825. *Corrections of the manuscript essay on the manufacture of tin-plate by Samuel Parkes.* [between 1785 and 1825]. 1 item (1 leaf)

Holograph ms. prepared "For the next volume of the *Memoirs of the Literary & Philosophical Society of Manchester.*" MSS 1124 A

367. Rennie, John, 1761-1821. *Letters.* 1785-1816. 3 items.

A.L.S. (1785 June 25, London) to John Scale concerning plans for a mill, A.L.S. (1810 Feb. 12, Ramsgate) concerning a conversation with Pond, and A.L.S. (1816 Oct. 8, Hotel de Libre, Rue du Helder). MSS 1204 A

368. Robbins, E. H., 18th cent. *Letter.* 1785. 1 item (2 p.)

A.L.S. (1785 Aug. 25, Boston) to Bourne concerning their search for a man named Hope Davis. MSS 1219 A

369. Delambre, Jean Baptiste, 1749-1822. *Letters.* 1786-1818. 9 items.

Nine A.L.S., some on Institut de France letterhead, to various correspondents including Ampère, Boissonade, and Dulong; in French. Additional material includes transcriptions. MSS 420 A

370. Lorenz, Johann Friedrich, 1738-1807. *Letters.* 1786-1797. 2 items.

A.L.S. (1786 Nov. 5, Koepenik) to Goeschen regarding publication of a book, and A.L.S. (1797 Sept. 7, Klosterberg) dealing with the publication of his translation of Euclid's work; in German. MSS 923 A

371. *[Notes on surveying].* [between 1786 and 1800]. 14 items: ill.

Notes and calculations on surveying (38 p., 7 leaves) citing Love's *Geodaesia,* 1 A.L.S. (1796 Aug. [11?] Cobham) concerning some trees with calculations on verso, and 2 surveying charts sewn in printed work. Laid in Hood's "Tables of difference of latitude and departure for navigators, land surveyors, etc. . . ," 1772.

MSS 1698 B

372. **Black, Joseph, 1728-1799.** *Letter.* 1787. 1 item (1 p.)

A.L.S. (1787 April 19, Edinburgh) recommending M. Girard. MSS 117 A

373. **Marti y Tranqués, Antonio, 1750-1832.** *Physica.* 1787-1788. [1], 373 [i.e., 375], [13] p.: ill.; 22 cm.

Treatise on the laws and nature of solid and fluid bodies, astronomy, optics, light, and color with tables; in Latin. Ms. ends with a poem in English: Thompson's "Autumn." Bound with two printed texts: *Theses mathematicas* by Joannes Delmas with Marti as arbiter, Paris: Seguy-Thiboust, 1788; and *Elementa matheseos.*

MSS 947 B

374. **Saussure, Nicholas Théodore de, 1767-1845.** *Letters.* [between 1787 and 1845]. 2 items.

A.L.S. to M. Guillaume Antoine de Luc congratulating him on his election, and A.L.S. to a friend about the summer weather; in French. MSS 1327 A

375. **Boehm, Andreas, 1720-1790.** *Letter.* 1788. 1 item (4 p.)

A.L.S. (1788 April 23, Giessen) recommending Flatt for Boehm's chair at the University, because of Flatt's sound theological views; in German. MSS 128 A

376. **Bramah, Joseph, 1749-1814.** *Papers.* 1788-1797. 2 items.

A.L.S. (1797 Nov. 13, Piccadilly) to Richard Neave & Co. concerning an order of hay and fodder, and a holograph petition for a patent on metal pipe manufacturing

with official letters of transmittal and recommendation (1788 Oct. 7-22). Additional material concerning Bramah's company. MSS 164 A

377. **Cooper, Astley Paston, Sir, 1768-1841.** *Letter.* [between 1788 and 1841]. 1 item (2 p.)

A.L.S. to R. Abraham concerning the best treatment for scrofula. MSS 369 A

378. **Haüy, René Just, 1743-1822.** *Correspondence.* 1788-1820. 8 items.

D.S. (1820 April 13, Paris) to Lamouroux also signed by Vauquelin and Desfontaines, 4 A.L.S. to Dutour, and 3 A.L.S. to Haüy from Dutour; in French. MSS 674 A

379. **Lavoisier, Antoine Laurent, 1743-1794.** *Papers.* 1788-1791. 4 items.

Three A.L.S. and an A. ms. leaf with S. attached; in French. MSS 860 A

380. **Boulton & Watt.** *Letter.* 1789. 1 item (2 p.)

A.L.S. (1789 Nov. 25, Birmingham) to Torres arranging a tour of a mill; in French with a Spanish notation in another hand. MSS 156 A

381. *Magnatismo: le sue capacità arrivono a tutte le relazione della vita--E assai in di la dei limiti del nostro picciolo globo.* 1789-1796. [158] leaves; 30 cm.

First page identifies work as "Lettere al signore de Heidenstam, Ambasciatore per il re di Suetia in Constantinopoli--6 Novembre 1789." Notes on next page: "Primo giorno dell Anno Magnetico--Febraro 6, 1795." Ms. has 57 numbered parts; part 51 begins: "5 Febraro 1796, Ultimo giorno dell'primo anno del Viaggiator divino." The text includes dialogue and verse; in Italian. MSS 858 B

382. **Académie des sciences (France).** *Abrégé des descriptions des artes par l'Académie des sciences.* [179-?]. 377 p.

A summary of the Académie's *Description des arts et métiers,* published 1761-1789. MSS 6 B

V. A letter of Galileo to Claude de Peiresc (12 May 1635).
Entry No. 58, SI Neg. 84-7194/5.

che io od altri volesse stampare, e che la proibizione è de omnibus editis, et edendis, si è presa assunto di fare stampare il resto delle mie fatiche non publicate ancora, e forse si è mosso p curiosità di veder l'esito di questa impresa, e che fortuna correranno tali materie lontaniss: da proposi: attenenti à religione più che non è il Cielo dalla Terra. Io contro à mia voglia sono stato forzato à concederne copia à S.A. sicuro che à me non ne possa succeder se non qualche travaglio, se bene non mi è stata fatta, ne accennata proibizione alcuna; p sò che non deuo ne anco hauer notizia del divieto fatto à gl'Inquisitori; p lo che questo che esseriuo à V.E. sia detto in confidenza. Da questo, e dall'esser stati raccolti in Firenze, et in Roma tutti l'opere mie si che più non se ne trouano p le librerie, apertam: si scorge che si fa ogni opera p leuar dal modo la mia memoria; nella qual vanità, se scopessero i miei auuersarij quanto poco io premo, forse non si moderebbero tanto ansiosi d'opprimermi.

Io non finirò di parlar cò lei senza di nuouo ringraziarla della sua infinita benignità, e del feruor col quale inuigila ne miei interessi, e se il solleuare chi fuor di tutte sue colpe vieni travagliato è atto meritorio, può V.E. vider sicura che ne ricevrà guiderdone dalla diuina bontà. E qui cò reuer: affetto gli bacio le mani, e nella sua buona grazia mi raccomando:

Dalla Villa d'Arcetri li 12 di Maggio 1635

Di V.S. Ill.ma et Ecc.ma

Deu. Et Obblig.mo ser:

Galileo Galilej

383. *Continuazione delle sostanze semplici metalliche.* [between 1790 and 1825]. 1 item (15 p.)

Manuscript treatise in Italian.   MSS 642 A

384.   G. F. *Della doppia distilazione di G. F.* [between 1790 and 1825]. 1 item (2 p.)

Manuscript; in Italian.   MSS 641 A

385.   Guglielmini, Giovanni Battista, 1764-1817. *Letters.* 1790-1791. 2 items.

Two A.L.S. (1790 Oct. 9 and 1791 July 16, Bologna) to Caldani, the former concerning printed costs; in Italian.   MSS 632 A

386.   Montgolfier, Joseph-Michel, 1740-1810. *Letter.* 1790. 1 item (2 p.)

A.L.S. (1790 April 21, Lamartre) concerning the actions of the Comité de Constitution; in French.   MSS 1034 A

387. *[Saponificazione].* [between 1790 and 1825]. 1 item (13 p.)

"Sezione II" of a work on saponification with the subtitle "Metodi practici di Fabbricazione"; in Italian.   MSS 643 A

388.   Schröter, Johann Hieronymus, 1745-1816. *Letter.* 1790. 1 item (4 p.)

A.L.S. (1790 Mar. 25, Lilienthal); in German.   MSS 1338 A

389.   Casanova, Giacomo, 1725-1798. *Letter.* 1791. 1 item (3 p.)

A.L.S. (1791 Feb. 4, Dux) to M. l'abbé D. Eusebio de la Lena, Vienna; in Italian.   MSS 309 A

390.   Dangos, Jean Auguste, 1744-1833. *Letter.* 1791. 1 item (2 p.)

A.L. (1791 July 16, Tarbes) to Messier concerning Dango's rejection of two job offers; in French.   MSS 399 A

391.   Chaptal de Chanteloup, Jean Antoine Claude, comte, 1756-1832. *Papers.* 1792-1823. 11 items.

Ten A.L.S. to various correspondents including Berthollet, Peuchet, and Molard, and 1 D.S.; in French.   MSS 333 A

392.   Chladni, Ernst Florens Friedrich, 1756-1827. *Papers.* 1792-1802. 6 items.

Four A.L.S. and 2 A. cards S.; in German and Latin.   MSS 342 A

393.   Desagneaux, P. C. L., 18th cent. *Letter.* 1792. 1 item (4 p.); ill.

A.L.S. (1792 Jan. 14, Crécy) concerning 2 inventions, a new barometer and steam propulsion for a boat; in French.   MSS 429 A

394.   Geoffroy Saint-Hilaire, Etienne, 1772-1844. *Mémoire sur l'anatomie comparée des organes électriques de la raie torpille, du gymnote engourdissant, et du silure trembleur par E. Geoffroy.* [between 1792 and 1802]. 15 leaves [i.e., 30 p.]; 27 cm.

Partial ms. (lacking about 300 words) of article published in *Annales du Muséum,* v. 1 (an XI, i.e., (1802); in French. Additional materials include photocopy of published article.   MSS 276 B

395.   Jussieu, Antoine Laurent de, 1748-1836. *Papers.* 1792-1800. 3 items.

Two A.L.S. (1792 Dec. 11, Paris; and an 8 messidor 6, i.e., 1800 June 25, Paris), and a holograph ms. concerning the work of Richard; in French.   MSS 764 A

396.   Marum, Martinus van, 1750-1837. *Dessein de la grande machine Teylerienne.* 1792. 1 leaf: ill.; 46 cm.

Drawing for improving Teyler's static electricity machine (1792 Oct. 8, Harlaam), marginal notes state that design was made for M. le Prince de Lambertini and is described in a letter to Landriani; in French. [XI]   MSS 911 B

397. *Synopis rei liquefactoriae generalis et particularis et principia metallurgiae, principaliores metallurgicorum [process--?] complectens manipulationes.* 1792. [526] p.: ill.; 40 cm.

Manuscript (1792 Mar. 21) signed illegibly, perhaps [A. Rillo?]. Topics include the fusion, liquefaction, and production of minerals, esp. silver, copper, iron, and zinc; in Latin. Spine title: Lectiones metallurgiae.

MSS 1289 B

398. **Vega, Georg, Freiherr von, 1754-1802.** *Letters.* 1792-1801. 4 items.

Four A.L.S.; in German.   MSS 1505 A

399. **Aldini, Giovanni, 1762-1834.** *Scientific papers.* 1793-1834. 2 boxes (ca. 80 items)

Manuscript notes entitled "Trattato sull'arte di segare il marmo"; ms. notes on animal electricity, galvanism, quarrying methods, fire prevention, and other experiments; correspondence with M. Halle and Prof. Maurice; printed pamphlets and articles; baptismal certificate; and documents conferring honors on Aldini. Finding aid, in Italian, with collection. Dealer correspondence with donor laid in. Box titles: [1] manuscripts -- [2] Invenzioni m.s.   MSS 18 B

400. **Bonpland, Aimé, 1773-1858.** *Letter.* [between 1793 and 1824]. 1 item (1 p.)

A.L.S. (undated) to Thouin; in French.   MSS 140 A

401. **Casali-Bentivoglia-Paleotti, Gregorio Filippo Maria, 1721-1802.** *Letters.* 1793-1794. 2 items.

Two A.L.S. [1793 April 23 and 1794 Feb. 18, Bologna) to Coltellini and M. le Comte Muran; in Italian.

MSS 308 A

402. **Cayley, George, 1773-1857.** *Letter.* [between 1793 and 1857]. 1 item (1 p.)

A.L.S. (undated) arranging for his unidentified correspondent to see Cayley's air engine.   MSS 317 A

403. **Fourcroy, Antoine François de, comte, 1755-1809.** *Papers.* 1793-1809. 11 items.

Seven A.L.S. to various correspondents including Ampère, Bonaparte, Pillet, Lanneau de Marey, and Larrey, 2 A.D.S., and 2 D.S.; in French.   MSS 531 A

404. **Fuss, Nikolai Ivanovich, 1755-1826.** *Letter.* 1793. 1 item (4 p.)

A.L.S. (1793 July 18, St. Petersburg) to Mechel concerning the state of arts and science in Russia; in French and German.   MSS 560 A

405. **Hassencamp, Johann Matthaus, 1743-1797.** *Letter.* 1793. 1 item (1 p.)

A.L.S. (1793 Mar. 1) to [Mossgraff?]; in German.

MSS 669 A

406. **Haüy, Valentin, 1745-1822.** *Letter.* 1793. 1 item (1 p.)

A.L.S. (1793 Jan. 2) recommending Antoine Auguste Senez for for a job at the Institution des Enfans-Aveugles; in French.   MSS 675 A

407. **Hube, Michael, 1737-1807.** *Letters.* 1793-1801. 8 items.

Eight A.L.S. to various correspondents; in German.

MSS 727 A

408. **Macculloch, John, 1773-1835.** *Letter.* [between 1793 and 1835]. 1 item (1 leaf)

A.L.S. (undated, Threadneedle Street) requesting prints of Cooke's view of Glen Tilt(?).   MSS 935 A

409. **Thilorier, Jean Charles, 1750-1818.** *Papers.* [between 1793 and 1818]. 3 items.

One A.L.S. to citoyen Mollere, 1 A.L.S. (an 6 Thermidor 16, i.e., 1798 July 16) to the Ministre de l'interieur concerning an invention, and ms. (3 p.): "Inventions et decouvertes [de Thilorier]"; in French.   MSS 1456 A

410. **Carnot, Lazare Nicolas Marquerite, comte, 1753-1823.** *Letters.* 1794-1807. 2 items.

Two A.L.S. (an 2 Germinal 13, i.e., 1794 April 2, and 1807); in French.   MSS 302 A

411. Lacépède, Bernard Germain Etienne de La Ville sur Illon, comte de, 1756-1825. *Papers.* 1794-1819. 34 items in two folders.

Twenty-two A.L.S., 8 D.S., and 4 A.D.S. to various correspondents including Chaptal de Chanteloup, Delattre, Desquiron, Huzard, and Jussieu; in French. Additional material in third folder. MSS 808 A

412. Leblanc, Nicolas, 1742-1806. *Papers.* 1794-1803. 4 items.

D.S. (an 3 vendémiaire 19, i.e., 1794 Oct. 10), 2 mss. concerning the manufacture of soda, and a certificate (an 11 ventôse 9, i.e., 1803 Feb. 28) presented to Leblanc by the Athénée des Arts; in French. MSS 862 A

413. Pearson, George, 1751-1828. *Observations sur une substance cireuse et au de la des chinois, ramassée a Madras par le Dr. Anderson, et qu' il a appelée lacque blanch par G. Pearson.* [1794?]. 1 item (7 p.) MSS 638 A

414. Gibbes, George Smith, Sir, 1771-1851. *Sur la conversion des substances animales en un matiere grasse qui rassemble beaucoup au blanc de baleine par G. S. Gibbes.* [1795?]. 1 item (2 p.)

Manuscript of article published in *Philosophical Transactions,* 1795; in French. MSS 637 A

415. Kitchiner, William, 1775?-1827. *Letters.* [between 1795 and 1827]. 2 items.

A.L.S. to Harley concerning the loan of the *Philosopher Hermit's Life,* and A.L.S. to Jerdan concerning Kitchiner's first book. Additional material includes several obituaries. MSS 785 A

416. Page, Frederick, 1769-1834. *The causes of the opposition of the Committee of Navigation of the City of London to the erection of pound lock & wiers by the Commissioners of the Upper Districts. . .* 1795. 16 p.: ill.; 22 cm.

Holograph ms. signed (1795 Apr. 14) with pencilled annotations, possibly in Rennie's hand. Bound with Rennie's "A report to the subscribers to a canal from Arundel to Portsmouth &c. &c. &c." printed in London (1816) but twice citing 1803 as a future date. MSS 1265 B

417. Vauquelin, Louis Nicolas, 1763-1829. *Letters.* 1795-1824. 7 items.

Three A.L.S. (an 4 Brumaire 21, i.e., 1795 Nov. 13, undated, and 1810 July 29) confirming events or giving character references, and 4 A.L.S. (1810-1824) to Geoffroy, Monsieur le Mercier, and Payen; in French. MSS 1504 A

418. Blumenbach, Johann Friedrich, 1752-1840. *Letters.* 1796-1821. 3 items.

Three A.L.S. to Cuvier, Deluc, and an unidentified correspondent; in French and German. Additional material includes translations. MSS 124 A

419. Germain, Sophie, 1776-1831. *Letter.* [between 1796 and 1831]. 1 item (1 p.)

A.L.S. concerning her admission to some assembly; in French. MSS 588 A

420. La Rive, Charles Gaspard de, 1770-1834. *Letters.* 1796-1805. 11 items.

Eleven A.L., most signed, to Marcet; in French. MSS 866 A

421. Laplace, Pierre Simon, marquis de, 1749-1827. *Papers.* 1796-1826. 16 items.

Ten A.L.S. and 5 D.S. to various correspondents including Chaptal de Chanteloup, Cuvier, Davy, and Oriani, one mentioning the Société Philomatique, and A. note of page corrections; in French. MSS 823 A

422. Monge, Gaspard, comte de Péluse, 1746-1818. *Letters.* 1796-1817. 5 items.

Four A.L.S., one mentioning Guyton de Morveau, and 1 L.S. dictated to his wife; in French. MSS 1025 A

423. Ritter, Johann Wilhelm, 1776-1810. *Papers.* [between 1796 and 1810]. 2 items.

A. notes S. (undated) on construction of "Voltaisches Batterie," and A. ms. (undated, 4 p.); in German. MSS 1218 A

424. Stratico, Simone, 1733-1824. *Papers.* 1796-1815. 3 items.

Two A.L.S. (1796 Aug. 26, Padua, and 1800 Aug. 4, Milan), and D.S. (1815 Aug. 18, Milan) concerning the

Cesarea Regia Censura and the [Regio] Istituto di Scienzie, Lettere ed Arti; in Italian. MSS 1424 A

425. Zach, Franz Xaver, Freiherr von, 1754-1832. *Papers.* 1796-1831. 3 items; ill.

One A.L.S. (1801 Aug. 1, Seeberg) in German, one A.L.S. (1831 Dec. 12, Paris) to a woman concerning astronomy and containing a table of predicted appearences of a comet in 1832, and a holograph leaf S. (1796 June 2, Seeberg) perhaps from a ledger, headed "118"; in French. MSS 1593 A

426. Darracq, François Balthazar, 1750-1808. *Letters.* 1797-1799. 2 items.

Two A.L.S., one (an 6 Brumaire 2, i.e., 1797 Oct. 23, Paris) to the Minister of War asking to accelerate the payment of an amount requested by the inhabitants of St. Jean Barbe, and one (an 7 Fructidor 27, i.e., 1799 Sept. 13, Paris) urging the signing of a document he submitted 3 weeks earlier; in French. MSS 404 A

427. Guyton de Morveau, Louis Bernard, baron, 1737-1816. *Papers.* 1797-1801. 2 items.

A.D.S. (an 6 frimaire 29, i.e., 1797 Dec. 19) attesting to his birthdate, and A.D.S. (an 9 prairial 23, i.e., 1801 June 12); in French. MSS 636 A

428. Hachette, Jean Nicolas Pierre, 1769-1834. *Papers.* 1797-1824. 3 items.

A.L.S. (1824 July 7, Paris) declining an invitation because of a scheduled exam, A.L.S. (Dec. 19) expressing pride at his appointment to professor, and D.S. (an 5 germinal 26, i.e., 1797 Mar. 25, Paris) "Rapport sur les solutions de Problêmes proposés pour le mois de Ventôse an 5."; in French. MSS 646 A

429. Pfaff, Johann Friedrich, 1765-1825. *Letter.* 1797. 1 item (4 p.)

A.L.S. (1797 July 17, Halmstedt) on mathematics; in German. MSS 1131 A

430. Rochon, Alexis Marie de, 1741-1817. *Letter.* 1797. 1 item (3 p.)

A.L.S. (an 5 pluviôse 14, i.e., 1797 Feb. 2, Brest) to an unnamed correspondent about the construction and supplying of a marine observatory, on stationary with the motto of the French Republic; in French. MSS 1227 A

431. Wilson, William, 18th cent. *[Mathematics].* 1797-1800. 71 [i.e., 140] p.: ill.; 20 cm.

Holograph ms. signed at least 4 times and dated periodically as Wilson completed sections of the text (1799 June 7-1800 Feb. 12). Topics include vulgar fractions, the rule of three, surds (i.e., irrational numbers), the square root, and the cube root. Master Vere de Vere has also signed two pages. Flyleaf: Wilson 1797. Cover: W. Wilson, His Book, 1799. Spine title: Math. MSS 1309 B

432. Clark, James, Sir, 1788-1870. *Letter.* [between 1798 and 1870]. 1 item (3 p.)

A.L.S., undated, to an unidentified female correspondent concerning a prescription for her headache. MSS 353 A

433. Gay-Lussac, Joseph Louis, 1778-1850. *Experiences sur les regles destinées à la mesure du base de l'arc terrestre.* [between 1798 and 1850]. [38] p.; 22 cm. MSS 232 B

434. Thornton, Robert Hohn, 1768?-1837. *Papers.* 1798-1813. 2 items.

One holograph leaf of receipts (1798 Feb. 6) of sums from Robert May for subscriptions to the *New Illustration of Linnaeus* for May and for Piozzi, and A.L.S. (1813 Aug. 13) to May concerning Thornton's lecture on botany. MSS 1465 A

435. Humboldt, Wilhelm, Freiherr von, 1767-1835. *Letters.* 1799-1829. 2 items.

A.L.S. (an 8 ce 5 me jour compl., i.e., Sept. 23, 1799?) to Fauris concerning studies in languages, in French; and A.L.S. (1829 July 27, Berlin) to Dr. Rosen concerning the grammar and writing of old languages; in German. MSS 735 A

436. Prony, Gaspard, Clair François Marie Riche, baron de, 1755-1839. *Papers.* 1799-1830. 5 items.

A.L.S. (1818 Feb. 1, Paris) concerning the "Fameux artist Bréquet, de l'académie Royal des Sciences," A.L.S. (1824 Feb. 5, Paris) concerning irrigation, A.L.S. (1830 Aug. 23, Paris) to M. Baude about the "Project d'endiquement du Rhone à Lyon," A.D.S. (le 28 brumaire an 8 de la rep. franc., i.e., 1799 Nov. 18) recommending Jean Francois(?) as a professor of mathematics, and A.D.S.

(1809 June 24, Paris) on a certificate form of the Ecole impérial des ponts et chaussées concerning Jean Baptîste Bertrand; in French. MSS 1176 A

437. Rumford, Benjamin, Graf von, 1753-1814. *Letters.* 1799-1806. 4 items.

A.L.S. (June 12) to Sir Joseph, two A.L.S. (1799 Aug. 5, 1800 April 8), and 1 A.L.S. (1806 March 12, Paris) in French to Seefeld, president of L'Académie Royale des Sciences et Arts, Munich. MSS 1458 A

438. Valperga di Caluso, Tommaso, 1737-1815. *Letters.* 1799-1812. 2 items.

A.L.S. (1799 Dec. 4, Turin), in Italian to Bodoni, a printer, concerning corrections in his book, and A.L.S. (1812 July 3, Turin), in French to M. Roatta, secretary to the Comté de Saint Martin, on behalf of his nephew; in French. MSS 1496 A

439. Brünings, Christian, 1736-1805. *Papers.* 1800. 2 items: ill.

A.L.S. (1800 March 13, Zwanerbring) to Eytelwein, with 1 p. of holograph calculations; in German. MSS 193 A

440. Cadet-de-Vaux, Antoine-Alexis, 1743-1828. *Letter.* 1800. 1 item (1 p.)

A.L.S. (le 7 frim. an 9; i.e., 1800 Nov. 28) to an unidentified correspondent concerning the Société de la Sciences and the *Journal* [*de Paris*?]; in French. MSS 291 A

441. *Dei corpi semplici e veduta nella lora influenza sulle [grandi!] operazioni della natura.* [between 1800 and 1840?]. 1 item (46 p.)

Manuscript (n.d.) on the chemical and physical properties of bodies, in six chapters. Topics include light, heat, and electricity; in Italian. MSS 1135 A

442. *Description de la pompe à feu.* [early 19th century]. 1 item (3 p.): ill.

Autograph ms.; in French. MSS 927 A

443. di Martino, O. *Letter.* [19th century?]. 1 item (1 leaf)

A.L.S. (Dec. 19); in Italian. MSS 967 A

444. Frankland, Thomas, Sir, 1750-1831. *Letter.* 1800. 1 item (1 p.)

A.L.S. (1800 April 29, London) to Smith regarding the excellence of *Flora Brittanica,* with 2 pages of later annotations by Ramsay. MSS 538 A

445. Guschi, Guido. *Descrizione d'un cammino, e stuffa di nuova invenzione del conte Guido Guschi.* [ca. 1830]. 31, [3] p.: folded ill.; 26 cm. MSS 280 B

446. Hahnemann, Samuel, 1755-1843. *Letter.* 1800. 1 item (2 p.)

A.L.S. (1800 Feb. 2, Altona); in German.

MSS 650 A

447. Lamarck, Jean Baptiste Pierre Antoine de Monet de, 1744-1829. *Letters.* 1800-1802. 2 items.

A.L.S. (an 8 pluviôse 19, i.e., 1800 Jan. 8) to Citoyen Agasse requesting money for Lamarck's illustrations for an encyclopedia, and A.L.S. (an 11 vendémiaire 16, i.e., 1802 Oct. 8, Paris) to an unnamed correspondent concerning meteorological observations; in French. MSS 815 A

448. Leycester, Robert. *Electricity [and] magnetism.* [19th century]. 124 p.: ill.

Holograph, signed on p. [1]. The article on electricity includes accounts of several experiments, including those of Franklin, Banks, Priestley, Volta, and Galvani. Article on magnetism notes variations reported in London in 1576, 1657, and 1776. Ten pen and wash drawings. 190 blank pages follow text. MSS 826 B

449. *Liste des muscles.* [19th century]. 1 item (5 p.)

A. ms. list; in French. MSS 1053 A

450. Massimo, Mario. *Papers.* [first half of 19th century]. 4 items.

Three A.L., two to Bellani, and an attached price list of scientific apparatus; in Italian. MSS 973 A

451. *[Mesmerismo trattato della pratrica della magnetizzazione].* [between 1800 and 1850]. 1 item (30 p.)   MSS 994 A

452. *Of the mensuration of solids.* 1800-1850. [176] p.: ill.; 26 cm.

Topics include "cylindric rings," munitions, specific gravity, and especially construction, e.g., the measurement of roofs, slate, timber, brick, glass. With definitions, rules, problems, tables, and examples.   MSS 1101 B

453. Prout, T. J. *Letter.* [19th century?]. 1 item (2 p.)

A.L.S. (May 9) to F. Symonds sending him "a piece of my late Father's writing." (Note: no writing sample now enclosed.)   MSS 1177 A

454. Smyth, W. *Letter.* [19th century]. 1 item (1 leaf)

A.L.S. to Rathbone thanking him for a book.   MSS 1384 A

455. Trento, Emilio. *Monografia del carbonio e dei suoi ossidi.* [1850?]. 216 p.; 21 cm.

Bound ms. concerning diamonds, graphite, "carbonio amorso," charcoal, "carbone animale," "carboni naturali," and various oxides of carbon; in Italian. Spine title: Carbonia   MSS 1301 B

456. Veneziani, Giuseppe, 1772-1853. *Trattato d'ottica dell'Abbate Giuseppe Veneziani, Professore di Fisica-Mathematica.* [19th century]. [6], 421 p., 23 leaves of plates: col. ill.; 20 cm.

Bound ms. containing treatise on optics. Illustrations include tables within the text and 91 figures in red and black ink on 23 folded plates at end of ms.; in Italian.   MSS 1305 B

457. Withering, William, 1775-1832. *Letter.* 1800. 1 item (1 leaf)

A.L.S. (1800 May 31, The Larches) to Robinson reasserting a claim for 100 pound sterling.   MSS 1579 A

458. Dalrymple, John, Sir, 1726-1810. *Petition.* 1801. 1 item (5 p.)

Holograph rough draft of a petition to the House of Commons (1801 July 25) concerning further experiments with his invention of soap made from fish.   MSS 393 A

459. Klaproth, Martin Heinrich, 1743-1817. *Letters.* 1801-1812. 3 items.

Two A.L.S. (1801 Feb. 25, and 1812 Jun. 1) in German, and A.L.S. (1811 Sept. 22, Berlin) to Vauquelin concerning an analysis of Tellure(?), in French.   MSS 786 A

460. Méchain, Pierre Francis André, 1744-1804. *Document.* 1801. 1 item (1 leaf)

A.D.S. (an 9 Pluviôse 10, i.e., 1801 Jan. 30) certifying and recommending Citoyen Grou; in French.
MSS 984 A

461. Pictet, Marc Auguste, 1752-1825. *Letters.* 1801-1825. 6 items.

Six A.L.S., three (all 1801, Paris) to Dr. Marcet, including one about officially replicating Volta's experiment; in French.   MSS 1142 A

462. Poisson, Siméon-Denis, 1781-1840. *Papers.* [between 1801 and 1840]. 2 items.

A.L.S. (1819 July 21, Paris) to Frullani about equations and an A. note S. (undated); in French.   MSS 1156 A

463. Siguad de La Fond, Joseph Aignan, 1730-1810. *Letters.* 1801-1804. 2 items.

A.L.S. (an 9 thermidor 16, i.e., 1801 Aug. 4, Bourges) to M. [Perboud?] concerning the Commission des poids et mesures, and A.L.S. (an 13 frimaire 14, i.e., 1804 Dec. 5, Bourges) to M. Haudry; in French.   MSS 1371 A

464. Struve, Friedrich Adolph August, 1781-1840. *Papers.* [between 1801 and 1840]. 4 items.

Two A.L.S., one concerning Prof. Muller and the [*Annals of Philosophy*], A.L.S. (1826 May 31) to Döbereiner, and a holograph leaf (1831 Dec. 21) giving a tally of receipts; in German.   MSS 1430 A

465. Werner, Abraham Gottlob, 1749-1817. *Letter.* [between 1801 and 1817]. 1 item (1 leaf)

A.L.S. to Breithaupt (perhaps Johann Breithaupt); in German. MSS 1554 A

466. Camper, Adriaan Gilles, 1759-1820. *Letter.* 1802. 1 item (3 p.)

A.L.S. (1802 nivôse 18, i.e., Jan. 8, Francker) to Cuvier concerning Cuvier's research on respiratory organs; in French. MSS 295 A

467. Corvisart des Marets, Jean Nicolas, baron, 1755-1821. *Document.* 1802. 1 item (1 p.)

A.D.S. (an 11 vendémaire 26, i.e., 1802 Oct. 18, Paris) requesting a position for citizen Leterisseur(?), signed also by Brisson, Deyeux, Haüy, and Lalande; in French. MSS 375 A

468. Dillwyn, Lewis Weston, 1778-1855. *Letters.* 1802-1826. 2 items.

A.L.S. (1802 April 20, Yarmouth) to Mr. Asquith concerning the completing of drawings and descriptions, A.L.S. (1826 Feb. 8, Penllergace) to Davy recommending to the Royal Society W. Osler's account of marine borers. MSS 442 A

469. Giobert, Giovanni Antonio, 1761-1834. *Papers.* 1802-1807. 10 items.

Manuscript articles on various chemical and mineralogical subjects, 5 of them in Giobert's hand; in Italian. Dealer's descriptive listing laid in. MSS 270 B

470. Legendre, Adrien Marie, 1752-1833. *Letters.* 1802-1824. 3 items.

Three A.L.S., one to Kératry, and one to a representative of Firmin-Didot; in French. MSS 871 A

471. Murphy, Patrick, 1782-1847. *Preface to the Weather Almanac.* [between 1802 and 1847]. 1 item (4 p.) MSS 1050 A

472. Seebeck, Thomas Johann, 1770-1831. *Letters.* 1802. 2 items.

Two A.L.S. (1802 May 28 and Apr. 20, [Bayreuth?]) to Rolf Becker; in German. MSS 1349 A

473. Cuvier, Georges, baron, 1769-1832. *Papers.* 1803-1832. 21 items.

Twelve A.L.S. to various correspondents including Delambre, Duméril, Prony, and Wiegman, 7 A.D.S. including 1 addressed to Van Rensselaer, 1 autograph card, and 1 page of notes; in French. Additional material includes translations and photocopies. MSS 390 A

474. Davy, Humphry, Sir, 1778-1829. *Papers.* 1803-1825. 14 items; port.

Nine A.L.S. to various correspondents including Ampère and Phillips, 2 holograph poems, a holograph scrap, a D.S. addressed to Hutton, and a portrait. Additional material includes A.L.S. from G. Curry to Marcet concerning Davy. MSS 411 A

475. Eytelwein, Johann Albert, 1764-1848. *Letters.* 1803-1823. 3 items.

Three A.L.S. (1803 April 25, 1814 Oct. 6, and 1823 Oct. 25); in German. MSS 496 A

476. Lefèvre-Gineau, Louis, 1751-1829. *Letters.* 1803-1822. 3 items.

Three A.L.S., two to Jussieu concerning bringing a poor woman into a hospital; in French. MSS 870 A

477. Pasquich, Johann, 1753-1829. *Document.* 1803. 1 item (1 leaf)

A.D.S. (1803 Dec. 4) headed "Contract" concerning publication of a book on mathematics by Schaumburg; in German. MSS 1099 A

478. Piazzi, Giuseppe, 1746-1826. *Letter.* 1803. 1 item (4 p.)

A.L.S. (1803 Dec. 2, Palermo) to Cagnoli, Presidente della scienze, on errors in Piazzi's catalogue; in Italian. MSS 1137 A

479. Bougainville, Louis Antoine de, comte, 1729-1811. *Letter.* 1804. 1 item (1 p.)

A.L.S. (an 12 pluviôse 30, i.e., 1804 Feb. 20) to Real concerning justice; in French. MSS 154 A

480. Crooke, Charles C. *Letter.* 1804. 1 item (11 p.); col. ill.

A.L.S. (1804 Dec. 8, Guernsey) containing a holograph copy of an earlier letter (1803 Sept. 20) to Lord St. Vincent concerning a method of enabling dismasted ships to bow the sea safely.   MSS 382 A

481. Vassalli-Eandi, Antonio Maria, 1761-1825. *Letters.* 1804-1822. 4 items.

Two A.L.S. (an 12 pluviôse 17, i.e., 1804 Feb. 7, and an 13 messidor 13, i.e., 1805 July 1, Turin) in French to Morozzo and Malacarne, and 2 A.L.S. (1817 Feb. 5 and 1822 Aug. 26, Turin) in Italian to Branchi and Fea.   MSS 1502 A

482. Benzenberg, Johann Friedrich, 1777-1846. *Papers.* 1805-1845. 6 items.

Five A.L.S. and 1 holograph ms. signed (15 p.); in German.   MSS 90 A

483. Adams, Dudley, fl. 1806-1814. *Correspondence.* 1806-1810. 11 items: ill.

Eight A.L.S. from Adams to Deluc and 3 holograph copies of letters from Deluc to Adams concerning the scientific apparatus Adams was making for Deluc's experiments.   MSS 63 A

484. Berthollett, Claude Louis, comte, 1748-1822. *Papers.* 1806-1816. 12 items: port.

Eleven A.L.S. to various correspondents including Berthier, Laplace, and Munge; and a roster of authors of the *Annales des sciences* (1816 March 17) with the signatures of Arago, Arcet, Chaptal, Gay-Lussac, and others; in French.   MSS 103 A

485. Dalton, John, 1766-1844. *Papers.* 1806-1831. 3 items.

A.L.S., D.S., and holograph leaf.   MSS 394 A

486. Fourier, Jean Baptiste Joseph, baron, 1768-1830. *Papers.* 1806-1829. 16 items.

Four D.S., 10 A.L.S. to various correspondents including Moreau de Jonnès, a printed scrap of text with marginal annotations, and a letter to Fourier; in French.   MSS 532 A

487. New York State Senate. *Document.* 1806. 1 item (1 leaf)

A resolution (1806 Jan. 30) to form a joint committee to propose a Botanic Garden. Signed by H. I. Blacker, Clerk.   MSS 1068 A

488. Playfair, John, 1748-1819. *Letter.* 1806. 1 item (1 leaf)

A.L. (1806 July 4, London) requesting that copies of his *Illustrations of the Huttonian Theory* be sent to the Royal Society, the Society at Manchester, and Pinkerton.   MSS 1147 A

489. Allen, William, 1770-1843. *Letter.* 1807. 1 item (2 p.)

A.L.S. (1807 Aug. 16, Penrith) to Deluc concerning electricity, to which is added a postscript by Howard on evaporation.   MSS 74 A

490. Bellani, Angelo, 1776-1852. *Letters.* 1807-1840. 12 items.

Twelve A.L.S., mostly to Massimo, including a 4 p. letter on instruments, and one to Gazzeri; in Italian.   MSS 89 A

491. Brongniart, Alexandre, 1770-1847. *Letters.* 1807-1846. 10 items.

Nine A.L.S. to various correspondents including Quérard, and Schimper, and a signed expense sheet; in French.   MSS 187 A

492. Conybeare, William Daniel, 1787-1857. *Letter.* [between 1807 and 1828]. 1 item (3 p.)

A.L.S. to W. Phillips concerning Phillips' plan to visit the Snowdon district.   MSS 367 A

493. David, Alois Martin, 1757-1836. *Letters.* 1807-1832. 2 items; ill.

Two A.L.S., one (1807 Dec. 10, Prag) concerning the repair of astronomical instruments and observation of a comet, and one (1832 Feb. 28, Prag) to Biela concerning comets; in German.   MSS 410 A

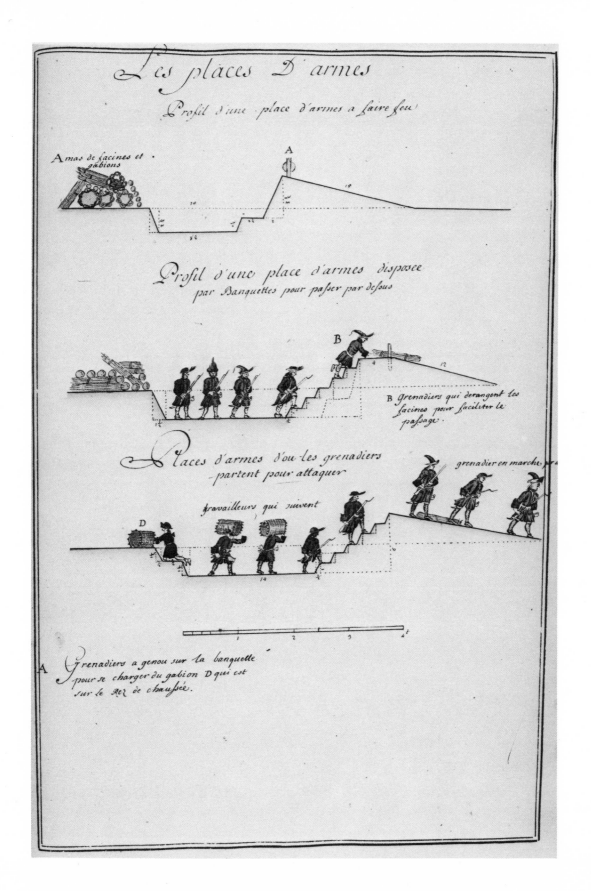

VI. Illustrations from an 18th century military *Tratté des siéges et de l'attaque des places.*
Entry No. 68, SI Neg. 82-4773.

494. Fulton, Robert, 1765-1815. *Letter.* 1807. 1 item (1 p.)

A.L.S. (1807 Feb. 13, Philadelphia) planning an experiment for May.   MSS 556 A

495. Gay-Lussac, Joseph Louis, 1778-1850. *Letters.* 1807-1843. 9 items.

Nine A.L.S. to various correspondents including Berthier, Chamberet, Coulier, and Darcet; in French.   MSS 578 A

496. Gehlen, Adolph Ferdinand, 1775-1815. *Letters.* 1807-1815. 2 items.

A.L.S. (1807 June 28, Halle), and A.L.S. (1815 Feb. 12, Munich) to Martius; in German.   MSS 580 A

497. Playfair, John, 1748-1819. *Notes on Natural Philosophy taken from Mr. Playfair's class during the winter session at Edinburgh.* 1807-1808. [240] p.: ill. (some col.); 21 cm.

Manuscript with "Rennie" penciled on inside cover and "years 1807 & 1808" penciled on title page; topics include motion and mechanics. Spine: George Rennie / Note Book / 1808.   MSS 1255 B

498. Smith, William, 1769-1839. *Papers.* 1807-1839. 13 items: ill.

Three A.L.S. (1816) to Sowerby, 3 A.L.S. (1816-1839) from Sowerby, J. Farey, and H. Jermyn, and ms. copy of L. to Jermyn, 2 leaves of illustrations of fossils by Sowerby, one annotated, his color pattern, his "Proposal for prints. . .," and two mss.: "Situations of Strata upon the Brighton Road. 1807. [Jm?] Farey [Senior] Esq.," and "Measurement of Strata found in [Boreing?]. . .," all pertaining to Smith's multi-volume work, *A Stratigraphical System of Organized Fossils.* Additional material includes 2 photographs of Smith's homes.   MSS 1382 A

499. Valmont de Bomare, Jacques Christophe, 1731-1807. *Letter.* 1807. 1 item (1 leaf)

A.L.S. (1807 May 22, Morsan) to Monsieur Le Danois, Membre du Corps Legislatif, concerning a sale of local cloth; in French.   MSS 1495 A

500. Walker, George, 1734?-1807. *Elements of geometry by the late Rev. George Walker, F.R.S., &c, &c, &c., transcribed from the author's M.S. by James Yates.* 1807. [155] p., 42, [3] leaves: ill.; 26 cm.

Bound holograph ms. signed "James Yates" on p. [155], containing text in 7 parts: Books I-III, Appendices to Books I-III, and Book IV are on pages 1-154 and Books V-IV and "Trigonometry" are on separately numbered leaves. 10 leaves of folded plates of diagrams are included. Signature on flyleaf: J. H. Lupton. Spine title: Geometry.   MSS 1306 B

501. Beddoes, Thomas, 1760-1808. *Letter.* 1808. 1 item (1 folded leaf)

A.L.S. (1808 Aug. 4, Eastbourn) to Phillips concerning an article on Isaac Jenkins.   MSS 82 A

502. Dierbach, Johann Heinrich, 1788-1845. *Note.* [between 1808 and 1845]. 1 item (1 p.)

A. note S. titled "In diesem Curse werde ich lesen"; in German.   MSS 440 A

503. Henry, William, 1774-1836. *Letters.* 1808-1832. 4 items.

Four A.L.S. to various correspondents including Faraday and Thomson.   MSS 687 A

504. Marum, Martinus van, 1750-1837. *Letter.* 1808. 1 item (3 p.)

A.L.S. (1808 Sept. 26, Haarlem) to an unnamed correspondent concerning an earlier note; in French.   MSS 968 A

505. Morse, Jedidiah, 1761-1826. *Letter.* 1808. 1 item (2 p.)

A.L.S. (1808 June 8, Boston) to Lyman concerning church affairs.   MSS 1040 A

506. Stanhope, Charles Stanhope, 3rd earl, 1753-1816. *Letter.* 1808. 1 item (1 leaf)

A.L.S. (1808 Jan. 18, Stratford Place) to Mr. Hugh Beams, Secretary of the London Vaccine Institution, declining with gratitude the favor proposed by the Board of Managers of the Institution.   MSS 1403 A

507. Thénard, Louis Jacques, baron, 1777-1857. *Letters.* 1808-1857. 9 items.

Seven A.L.S. to various correspondents including Civiale. Bouv-[Bourdier?], and Sénarmont, and 2 A.L.S. testifying to the work of students; in French.　MSS 1454 A

508. Chappe, Ignace Urbain Jean, 1760-1828. *Letters.* 1809-1821. 3 items.

Three A.L.S. concerning the mechanical telegraph; in French.　MSS 332 A

509. Couch, Jonathan, 1789-1870. *Notes on the habits of British stalk eyed crustacean animals.* [between 1809 and 1870]. [47] leaves: col. ill.: 35 cm.

Two watercolor ill. laid in at leaf [39]; a half-sheet of text laid in at leaf [41]. With: "Cornish Fauna" (148 p.) dated 1857; and 2 other holographs on natural history totaling 33 leaves.　MSS 234 B

510. *Diverse memorie ed osservazioni chimiche e fisiologiche sull'Ipecacuana sopra i di Lei succedanei e sopra alcune nuove sue preparazioni.* 1809-1817. 79 p.; 23 cm.

Articles and notes on ipecac, its substitutes, and new preparations for it, published in the *Bulletin* and the *Journal de Pharmacie* of Paris between 1809 and 1817; translated into Italian by G. F.(?). Authors include P. A. Masson-Tour, T. L. A. Loiseleur Deslongchamps, and Vittore Poadein.　MSS 937 B

511. Malus, Etienne Louis, 1775-1812. *Letter.* 1809. 1 item (2 p.)

A.L.S. (1809 Feb. 20, Paris) to M. Drouet(?) concerning fortifications; in French.　MSS 952 A

512. Schumacher, Heinrich Christian, 1780-1850. *Papers.* 1809-1850. 11 items.

Four A.L.S. (1828-1847) in German, 5 A.L.S. (1820-1850) in French, one to Arago, one to Bouvard, a second to Bouvard written on a duplicated A.L.S. in Latin on "Elementa cometae" (1826), 1 to Chevalier, and 1 concerning Cuvier, 1 holograph ms. listing 22 books, and 1 autograph signature (1809 Nov. 12).　MSS 1340 A

513. Ciccolini, Lodovigo, 1767-1854. *Papers.* 1810-1830. 6 items.

Three A.L.S. (1810-1830) to Calandrelli and Poletti, A. list of poetry books, and 2 holograph articles signed; in Italian.　MSS 347 A

514. Forster, Benjamin Megget, 1764-1829. *Letters.* 1810-1812. 3 items.

Three A.L.S. to Deluc with information about experiments on electric bells.　MSS 529 A

515. Manton, Joseph, 1766?-1835. *Document.* 1810. 1 item (3 p.)

Recommendation (1810 Jan. 4) of V. Gibbs to the king concerning Joseph Manton's petition for a patent on a telescope.　MSS 955 A

516. Séguin, Armand, 1765-1835. *Letters.* 1810-1821. 2 items.

One A.L.S. (1810 Mar. 27) to Monsieur [Nichomme?], and A.L.S. (1821 Mar. 5) to the president of the Académie royale des sciences; in French.　MSS 1352 A

517. Smith, James Edward, Sir, 1759-1828. *Letter.* 1810. 1 item (3 p.)

A.L.S. (1810 March 23, Norwich) to Dr. Sims concerning a review of Smith's work in the *Edinburgh Review* and proposed scientific names, i.e., Conchium, Linnaeus, and Linnaen.　MSS 1381 A

518. Ure, Andrew, 1778-1857. *Letters.* 1810-1851. 5 items.

Four A.L.S., one with A. envelope, to various correspondents including West.　MSS 1491 A

519. Venturi, Giovanni Battista, 1746-1822. *Letters.* 1810-1819. 2 items.

A.L.S. (1810 Dec. 8, Bern) to Arnaldi, Segretario dell 'Istituto Nazionale a Bologna, on letterhead of Regno d'Italia, L'Agente Diplomatico nella Swizzera, concerning the writer's study of minerals, A.L.S. (1819 Oct. 12, Reggio) to Molini concerning Galileo's works; in Italian.　MSS 1506 A

520.   Whitney, Eli, 1765-1825. *Papers.* 1810-1814. 5 items: ill.

Four A.L.S. (1813-1814, New Haven) to Amasa Davis, Irvine, and Deceus Woosworth, and a pen and pencil drawing of "Mr. Whitneys' Machine to prove gunpowder" (1810 Aug. 11).　　　　　MSS 1560 A

521.   Bode, Johann Elert, 1747-1826. *Letters.* 1811-1826. 4 items.

Four A.L.S. to various correspondents, including Biela and Gilbert; in German.　　　　　MSS 126 A

522.   Bostock, John, 1773-1846. *Letter.* 1811. 1 item (1 p.)

A.L.S. (1811 Feb. 3) thanking an unidentified lady for her elegant present.　　　　　MSS 152 A

523.   Bowdich, Thomas Edward, 1791-1824. *Letter.* [between 1811 and 1824]. 1 item (1 p.)

A.L.S. requesting the loan of a botanical book. Additional materials include a photocopy of a certificate from the Académie signed by Cuvier thanking Bowdich for his book on conchology (1822 July 29); in French.　　　　　MSS 160 A

524.   Gergonne, Joseph Diez, 1771-1859. *Letter.* 1811. 1 item (4 p.)

A.L.S. (1811 Sept. 24, Nîmes) to M. Dhombres Firmas explaining why Gergonne could not visit him, and commenting on two papers on polarization optics which lack conclusive proof; in French.　　　　　MSS 586 A

525.   Humboldt, Alexander von, 1769-1859. *Letters.* 1811-1858. 24 items: ill., ports.

Twenty-one A.L.S. to various correspondents including Schumacher and Milne-Edwards, 2 envelopes, and photograph; in German and French.　　　　　MSS 734 A

526.   *Rapport sur le procès de Galilée et commencement de ce procès.* 1811. 29 p.; 23 cm.

Manuscript with introductory letter (1811 Mar. 23, Paris) to the Emperor proposing to translate the trial of Galileo and mentioning "la traduction littérale des premieres pièces" which follows; these pieces are "Contre Galileo Galilei" and transcription of 2 letters including one from Galileo to Benedetto Castello; on paper water-marked with the image of "Napoléon le Grand"; in French.　　　　　MSS 1262 B

527.   Wollaston, William Hyde, 1766-1828. *Papers.* 1811-1822. 4 items: ill.

One A.L.S. (1814 July 16, Fitzroy) to Bostock introducing Gamel', 1 A.L.S. declining a dinner invitation, one sketch of an electro-magnet by Wollaston for Dr. Dalton (1822), and A.L.S. (1811 Dec. 5) from Marcet to Wollaston.　　　　　MSS 1584 A

528.   Ampère, André Marie, 1775-1836. *Letters.* 1812-1835. 14 items.

Thirteen A.L.S. in French to various correspondents including Bouchu, Bredin, Daubrée, and Maine de Biran; and ms. copy of letter to Biran. Accompanied by photocopies.　　　　　MSS 76 A

529.   Bolzano, Bernard, 1781-1848. *Papers.* 1812-1836. 2 items.

A.L.S. (1836 July 18) in German; and D.S. (1812 Dec. 22) for Karesoh Wenzeslaus in Latin.　　　　　MSS 138 A

530.   Farrar, John, 1779-1853. *Letter.* 1812. 1 item (4 p.)

A.L.S. (1812 Oct. 29, Cambridge) to Cleveland concerning electric batteries.　　　　　MSS 501 A

531.   Franklin, F. W. *Letter.* 1812. 1 item (1 p.)

A.L.S. (1812 April 20, Hartford) to Verner and Hood, London, requesting a copy of "The Universal Guide to the stream of time," on approval.　　　　　MSS 540 A

532.   Gauss, Karl Friedrich, 1777-1855. *Papers.* 1812-1840. 12 items.

Ten A.L.S., and 2 D.S.; in German. Additional material in folder.　　　　　MSS 575 A

533.   Larrey, Dominique Jean, baron, 1766-1842. *Letters.* 1812-1813. 3 items.

Three A.L.S., one to David; in French.　　　　　MSS 842 A

534. *Rapport sur les mines de Sambel et Chessy.* [between 1812 and 1850?]. 1 item (4 p.)

A. ms. (undated) of a report mentioning mine accidents, e.g., one in 1811, with numerous revisions of the text; in French.   MSS 1189 A

535. Berzelius, Jöns Jakob, Freiherr, 1779-1848. *Papers.* 1813-1847. 8 items: port.

Seven A.L.S. (in Swedish, French, and German) to various correspondents including Olivier and Sochez; holograph ms. entitled "Analytische Untersuchung des Bitterwassers von Saidschutz in Böhmen" (10 p.); in German.   MSS 105 A

536. Giovio, Giovanni Battista, conte, 1748-1814. *Papers.* 1813. 2 items.

A.L.S. (1813 Sept. 13, [Verzgo?]) to Pizzi with enclosure of a ms.: "Dell'azione magnetica della terra"; in Italian.   MSS 594 A

537. Lardner, Dionysius, 1793-1859. *Letters.* [between 1813 and 1859]. 6 items.

Six A.L.S., only one dated (1830 Mar. 8, London) to Montgomery.   MSS 825 A

538. Makdougall-Brisbane, Thomas, Sir, 1773-1860. *Letters.* 1813-[1842?]. 2 items.

Two A.L.S., one (1813 Oct. 17, Spain) to Sir Joseph Banks concerning a hail-storm and one (1842? April 18, Makerstoun) to Edward Milne, concerning Milne's article on the oscillations of the sea.   MSS 1234 A

539. Ross, John, Sir, 1777-1856. *Papers.* 1813-1826. 2 items.

A.L.S., (1826 Mar. 23, South Kilworth) to Rev. Samuel Fennell, and D.S. (1813 Mar. 31) concerning the wages due William Burgess, a carpenter on the *Briseis* under Captain Ross.   MSS 1234 A

540. Rossel, Elisabeth Paul Edouard de, 1765-1829. *Letter.* 1813. 1 item (1 leaf)

A.L.S. (1813 July 21, Paris) to de Vindé regretfully deferring his visit to the home of de Vindé; in French.   MSS 1236 A

541. Thomson, Thomas, 1773-1852. *Letter.* 1813. 1 item (1 leaf)

A.L.S. (1813 Oct. 29) to Mr. Knight Spencer of the Surry Institution concerning a proof and the apparatus for a series of lectures.   MSS 1463 A

542. Deluc, Jean André, 1727-1817. *Essai d'histoire de la physique, ou, des progrès qu'on fait dans les siècles précédens par J. A. Deluc.* 1814. [3], 239 p.; 27 cm.

Additional material includes an A.L.S. (1814 July, Windsor) to André Deluc, his nephew, explaining that the article is enclosed, and giving some information about J. A. Deluc's life and experiments; in French.   MSS 277 B

543. Döbereiner, Johann Wolfgang, 1780-1849. *Letters.* 1814-1841. 3 items.

Three A.L.S., one mentioning Ørsted; in German.   MSS 447 A

544. Gamel', Iosif Khristianovich, 1788-1861. *Letters.* 1814. 2 items.

A.L.S. (1814 Jan. 11, Bath) to Mr. Booth (bookseller) in German, and A.L.S. (1814 Nov. 24, Edinburgh) to Holt concerning the new educational system.   MSS 644 A

545. Lacroix, Silvester François, 1765-1843. *Letter.* 1814. 1 item (2 p.)

A.L.S. (1814 Feb. 7, Paris) to M. Monpence Michaud commenting on an article about the Roman mathematician and architect Anthemius; in French.   MSS 809 A

546. Leslie, John, Sir, 1766-1832. *Papers.* 1814-1827. 3 items.

A.L.S. (1814 April 30, Edinburgh) to Clarke introducing Napier, A. note to Miss Hullmandel, and admission ticket (1827 Dec. 1) to Lectures on Natural Philosophy, University of Edinburgh to Downie signed by Leslie.   MSS 885 A

547. Liston, Robert, 1794-1847. *Letter.* [between 1814 and 1847]. 1 item (1 leaf)

A.L.S. (undated) to an unidentified correspondent concerning an invitation to eat haggies.   MSS 914 A

548.   Moretti, Giuseppe, 1782-1853. *Letter.* [after August 1814]. 1 item (1 leaf)

A.L.S. (Milan) to Mme. Buynard ordering the book *Le Botanicon Parisiense* of M. Vaillant; in French.

MSS 1037 A

549.   Paris, John Ayrton, 1785-1856. *Letter.* 1814. 1 item (2 p.)

A.L.S. (1814 Nov. 17, Penzance) concerning engravings for a book.   MSS 1096 A

550.   Vaidy, Jean Vincent François, 1776-1830. *Letter.* 1814. 1 item (3 p.)

A.L.S. (1814 Nov. 24, Paris) to Monsieur le President (Banks?) concerning "l'eau medicinale" and "colchicum autumnale"; in French.   MSS 1493 A

551.   Arago, Dominique Fransçois Jean, 1786-1853. *Papers.* 1815-1850. 32 items.

Letters from Arago, including A.L.S. to Berthier, Milne-Edwards, Ørsted, an A.L.S. recommending Bruslé for a professorship, and an A.L.S. mentioning the *Annales de chimie;* a fragment of a eulogy for Volta; and A.D.S. acknowledging receipt of books from Lubbock, Wheatstone, and Huot in Arago's position as secrétaire perpétuel de l'Académie royale des sciences; in French and English. Facsimile of Arago's hand in folder.   MSS 19 A

552.   Biot, Jean Baptiste, 1774-1862. *Letters.* 1815-1856. 11 items: port.

Eleven A.L.S. to various correspondents including Bersin, Blagden, and Lubbock. Additional materials include photocopies.   MSS 116 A

553.   Brewster, David, Sir, 1781-1868. *Papers.* 1815-1865. 17 items.

Fifteen A.L.S. to various correspondents including one (1821 Mar. 24, Edinburgh) to Ørsted concerning Brewster's recommendation for Ørsted's membership in the Royal Society of Edinburgh; a signed inscription to one of his children; and an envelope.   MSS 176 A

554.   Brunel, Madame. *Letter.* 1815. 1 item (2 p.)

A.L.S. (1815 April 27, Chelsea) to the Duc de la Rochefoucauld, Paris, regretting that M. Brunel has just left and cannot answer the Duc's letter himself; in French.   MSS 191 A

555.   Fresnel, Augustin Jean, 1788-1827. *Papers.* 1815-1822. 4 items.

Two A.L.S. and 2 D.S.; in French.   MSS 546 A

556.   Herapath, John, 1798-1868. *Letter.* 1815. 1 item (1 p.)

A.L.S. (1815 July 16, Cranford) to the editor of the *Annals* saying that Herschel is not the author of an erroneous solution in Babbage's paper.   MSS 690 A

557.   Hildebrandt, Georg Friedrich, 1764-1816. *Letter.* 1815. 1 item (1 p.)

A.L.S. (1815 Dec. 2, Erlangen) returning a book of which he received 2 copies; in German.   MSS 702 A

558.   Hossfeld, Johann, 1768-1837. *Letter.* 1815. 1 item (1 p.)

A.L.S. (1815 April 18, Dreissigacker) urging an unnamed correspondent to submit Hossfeld's ms. dealing with the Attraction Theory to the Academy as promised or to send it back; in German.   MSS 725 A

559.   *Notes and observations on various subjects connected with artillery.* [1825?] 155 [i.e., 165], [34] p.: ill. (some col.), tables; 24 cm.

On making, using, and transporting the guns, mortar, ammunition, etc. of the British Army ca. 1825. Profusely illustrated with watercolors and ink diagrams, charts, and tables. Spine title: Artillery. With 170 blank pages after text. Dealer's description laid in.   MSS 1006 B

560.   Rennie, George, 1791-1866. *Journal of the year 1815.* 1815-1816. [160] p.; 20 cm.

Holograph diary signed (1815 Jan. 1 through 1816 Dec. 31). Cover: 1815 & 16 G. R.   MSS 1264 B

561.   Angerstein, John Julius, 1735?-1823. *Letter.* [1816?]. 1 item (1 leaf)

Dinner invitation to Mr. and Mrs. Pond and Miss Bradley, probably the daughter of James Bradley.   MSS 1611 A

562. Brunel, Marc Isambard, Sir, 1769-1849.
Papers. 1816-1843. 11 items.

Ten A.L.S. to various correspondents including Bréguet, D'Avannes, and Hudson, and one calligraphic sample; in English and French. Additional materials include photocopies and translations.   MSS 192 A

563. France. Ministère de la guerre. *Telegram.* 1816. 1 item (1 leaf)

Telegram (1816 Sept. 24, Paris) from M. Leleu, the Minister of War, to Bouthillier-Chavigny as Préfet du Bar Rhin authorizing a new credit of 80,000 [francs?]. Handwritten on engraved form and signed as correct by the director of the telegraph company; note added, perhaps by Bouthillier-Chavigny; in French.   MSS 1450 A

564. Galloway, Thomas, 1796-1851. *Letters.* [between 1816 and 1851]. 2 items.

Two A.L.S. (May 24 and June 25, Serjeants Inn) concerning a revised article and a magazine advertisement.   MSS 565 A

565. Jouffroy, Théodore Simon, 1796-1842. *Letter.* [between 1816 and 1842]. 1 item (3 p.)

A.L.S. (Dec. 18, Paris) to Marquis Cesar Boccella with thanks for the kind reception; in French.   MSS 760 A

566. Kries, Friedrich Christian, 1768-1849. *Papers.* 1816-1827. 3 items.

Two A.L.S. and a holograph ms.; in German.   MSS 800 A

567. Ørsted, Hans Christian, 1777-1851. *Papers.* 1816-1850. 18 items.

Thirteen A.L.S. to various correspondents including Arago, Carus, Dulong, Hansteen, Schubarth, Thomson, 2 A.S., 2 D.S., and 1 D. signed by Frederik; in Danish, French, and German.   MSS 1083 A

568. *Rapport sur une demande faite par le [--o?] Paganon d'être autorise a continuer le [deroulemans?] de son usine cottone. . .* 1816. 1 item (3 p.)

A. ms. report on Jean Claude Paganon's request (1816 Mar. 20) to be authorized to keep his factory in operation and on pertinent legal observations, esp. concerning the law of 21 April 1810; in French.   MSS 1190 A

569. Stuart, Charles, Sir, 1779-1845. *Letter.* 1816. 1 item (1 leaf)

Manuscript copy of an A.L.S. (1816 Sept. 8, Paris) to the Duke de Richelieu requesting permission for Rennie, Watt, and an interpreter to visit the French dockyards.   MSS 1432 A

570. Bessel, Friedrich Wilhelm, 1784-1846. *Letters.* 1817-1843. 6 items: port.

Six A.L.S. to various correspondents including Delambre and Utzschneider; in German and French.   MSS 106 A

571. Boissy d'Anglas, François Antoine, comte de, 1756-1826. *Letter.* 1817. 1 item (8 p.)

A.L.S. (1817 Dec. 2, Paris) concerning the Montgolfier brothers; in French.   MSS 135 A

572. Brandes, Heinrich Wilhelm, 1777-1834. *Letters.* 1817-1833. 2 items.

Two A.L.S. (1817 July 3 and 1833 Dec. 15), one to Configliachi; in German.   MSS 169 A

573. Dupin, Charles, baron, 1784-1873. *Letter.* 1817. 1 item (2 p.)

A.L.S. (1817 Jan. 5, London) to Prony concerning work and travel plans; in French.   MSS 464 A

574. Jussieu, Adrien de, 1797-1853. *Notice sur les travaux de botanique de M. A. DeJussieu.* [between 1817 and 1853]. 1 item (13 p.)   MSS 763 A

575. Long, Charles Edward, 1796-1861. *Letter.* [between 1817 and 1861]. 1 item (1 leaf)

A.L.S. (Mar. 31) to Jerdan, concerning the *Literary Gazette.*   MSS 920 A

576. Rhind, William, fl. 1833-1867. *[Lecture notes on the natural history of man].* [between 1817 and 1858]. 222 [i.e., 216] p.: ill.; 19 cm.

Manuscript of notes on natural history. Topics include clouds, geology, plants, and human anatomy. Text concludes: "Mr. Rhind concludes the present course of lectures on the natural history of man. . ." Spine title: Notes Nat. Hist. I. First 4 pages lacking. "1856" is written inside front cover. Watermark: Joseph Coles 1817.

MSS 1268 B

577. Scrope, George Julius Duncombe Poulett, 1797-1876. *Letter.* [1825?]. 1 item (3 p.)

A.L.S. (undated, probably 1825) to Mr. W. Phillips about a ms. with plates that Phillips is to publish for Scrope. MSS 1165 A

578. Sowerby, James, 1757-1822. *Letter.* 1817. 1 item (1 leaf)

A.L.S. (1817 Aug. 27, Lambeth) to Mr. Cobbold concerning a gift of crayfish to stock Mr. Cobbold's stream. MSS 1394 A

579. Wedgwood, Josiah, 1730-1795. *Letter.* 1817. 1 item (1 leaf)

A.L.S. (1817 June 18, Etruria) to James Pine concerning a new teaware pattern. MSS 1540 A

580. Bowditch, Nathaniel, 1773-1838. *Papers.* 1818-1829. 2 items.

A. check S. (1818 Dec. 9) and A.L.S. (1829 May 4, Boston) to Emerson concerning an honor to Bowditch from the Boston Mechanics Institute. MSS 159 A

581. Despretz, César Mansuète, 1789-1863. *Letters.* [between 1818 and 1823]. 4 items.

Four A.L.S. (one dated 1848 July 13) to Lormier, Moigno, and unnamed correspondents; in French. MSS 433 A

582. Faraday, Michael, 1791-1867. *Papers.* 1818-1865. 113 items in 4 folders: ill., port.

Two autograph signatures, 2 A. admittance tickets S., holograph notes (4 p.), 107 A.L.S. and 1 envelope to various correspondents including Barry, Brodie, Brunel, Codrington, Fox, Hawes, Lardner, Lyell, Melloni, Mount-Temple, Murchison, Tomlins, and Watson. Additional material in fifth folder. MSS 554 A

583. Gattinara, C. *Letter.* 1818. 1 item (1 p.)

A.L.S. (1818 Jan. 23, Turin) to Chiaveroti; in Italian. MSS 640 A

584. Gregory, Olinthus Gilbert, 1774-1841. *Letter.* 1818. 1 item (3 p.)

A.L.S. (1818 Aug. 11, Seaview near Ryde) to Montgomery concerning their friendship, their health, Gregory's holiday, and Montgomery's poetry. MSS 620 A

585. Hermbstädt, Sigismund Friedrich, 1760-1833. *Letter.* 1818. 1 item (1 p.)

A.L.S. (1818 Aug. 6, Berlin) to Jacquin; in German. MSS 691 A

586. Herschel, Caroline Lucretia, 1750-1848. *Paper.* [between 1818 and 1848]. 1 item (2 p.)

Holograph ms. anecdote of W. Herschel, copied from *The Pocket Magazine of Classic and Polite Literature,* vol. II [1818?]. MSS 693 A

587. Lampadius, Wilhelm August, 1772-1842. *Letters.* 1818-1833. 2 items.

A.L.S. (1818 Dec. 11, Freiberg) concerning an invitation to join a Test Association, and A.L.S. (1833 Aug. 25, Freiberg) concerning information about a strange kind of slag; in German. MSS 818 A

588. Madden, Richard Robert, 1798-1886. *Letters.* [1855?]. 2 items.

A.L.S. (1855 Sept. 25) to Mrs. Wilde concerning an invitation, and A.L.S. (undated but probably 1855 Sept. 27) to Wilde concerning a census. MSS 942 A

589. Vogel, Heinrich August, 1778-1867. *Papers.* 1818-1844. 4 items.

Two A.L.S. (1840, 1844, Munich) to Barth, one A.L.S. (1818 Apr. 2, Munich) to a professor, and a holograph ms. (4 p.) containing a short autobiography and list of professional accomplishments and memberships; in German. MSS 1511 A

590. Brande, William Thomas, 1788-1866. *Notes on a course of lectures on chemistry delivered by Mr. Brande in the Laboratory of the*

Royal Institution in 1819. 1819. [300] p.: ill.; 19 cm.

On spine: Brande - Notes on chemistry - 1819.

MSS 224 B

591. Clapeyron, Benoît Paul Emile, 1799-1864. *Vitesse des ondes.* [between 1819 and 1864]. 1 item (8 p.)

Holograph article signed; in French.　MSS 350 A

592. Dobbins, James. *Letters.* 1819. 2 items.

Two A.L.S. (1819 Aug. 29) to the Earl of Liverpool, presenting Dobbins' Magnetic Remarker and seeking employment.　MSS 446 A

593. Murray, John, 1786?-1851. *Letters.* 1819-1823. 2 items.

Two A.L.S. (1819 July 20 and 1823 May 21) both to the Editor, *Gentlemen's Magazine* (i.e., Nichols).

MSS 1063 A

594. Ampère, Jean Jacques, 1800-1864. *Letter.* [1835?]. 1 item (3 p.)

A.L.S. ([1835?] June 26) to Abrahams in French.

MSS 77 A

595. Hall, Basil, 1788-1844. *Magnetical observations made on board* H.M.S. Conway *at Portsmouth Harbour & on South American Station by Basil Hall & Henry Foster.* 1820-1822. [82] p.: charts; 20 cm.

Includes Barlow's remarks on the observations. Additional material in ms. folder.　MSS 221 B

596. Orioli, Francesco, 1785-1856. *Letter.* 1820. 1 item (2 p.)

A.L.S. (1820 Jan. 30, Bologna) to Gazzeri; in Italian.

MSS 1088 A

597. Phillips, Richard, 1778-1851. *Letter.* [between 1820 and 1851]. 1 item (1 leaf)

A.L.S. (Oct. 8, Nelson Square) to Mr. Thompson, a chemist, requesting the loan of "the Pharmacopia lately published in America."　MSS 1134 A

598. Sabine, Joseph, 1770-1837. *Letter.* 1820. 1 item (2 p.)

A.L.S. (1820 Sept. 25, Horticultural Society) to Mr. Shepherd.　MSS 1315 A

599. Scherer, Alexander Nicolaus, 1771-1824. *Letter.* 1820. 1 item (1 leaf)

A.L.S. (1820 Jan. 14, St. Petersburg) to an unknown correspondent announcing his election as an honorary member of the Imperial Pharmaceutical Society.

MSS 1329 A

600. Weiss, Christian Samuel, 1780-1856. *Papers.* 1820-1831. 2 items.

A.L.S. (1820 [Aug.?], Berlin) in German, and holograph leaf (1831 Nov. 6, Berlin) announcing in Latin his courses in minerology and crystallography.

MSS 1545 A

601. Wöhler, Friedrich, 1800-1882. *Papers.* 1820-1879. 18 items.

Eleven A.L.S. ([1820?]-1878, mostly from Göttingen) to various correspondents including Meyer, Ørsted, and Weber, 4 A.L.S. (1856-1879) in French to various correspondents including Weber and possibly Séarmont, one A. envelope S., one holograph leaf S.: "Sommer-Semester 1849," and one ms. leaf: "Dr. Friedrich Wöhler" perhaps written by Dr. F. Hausmann (1852 Oct. 6, Göttingen). Topics of the letters include zirconium, Weber, Leibig, and Fremy; in German.　MSS 1581 A

602. Bosc, Louis Augustin Guillaume, 1759-1828. *Letter.* 1821. 1 item (1 p.)

A.L.S. (1821 Jan. 25) to Marron concerning a meeting Bosc had promised to attend; in French.　MSS 148 A

603. Configliachi, Pietro, 1779-1844. *Letters.* 1821-1826. 2 items.

Two A.D.S. (1821 June 10, 1826 June 21) acknowledging receipt of money and a dissertation to be placed in the University Library; in Italian.　MSS 366 A

604. Deyeux, Nicolas, 1745-1837. *Letters.* 1821-1830. 2 items.

Two A.L.S. (1821 Nov. 7, 1830 Oct. 24, Paris) to Jussieu; in French.　MSS 437 A

VII.   A page from the *Vegetation of metals* of Isaac Newton
(ca. 1700).

605. Fraunhofer, Joseph von, 1787-1826. *Papers.* 1821. 2 items.

A.D. (1821 Aug. 2), and holograph ms. S. "Beschreibung eines neuen Mikrometers" (13 p.); in German.
MSS 543 A

606. Geoffroy Saint-Hilaire, Etienne, 1772-1844. *Papers.* 1821-1838. 10 items.

Eight A.L.S. to various correspondents including Andouin, Arago, Chabrol, and Prunelle, holograph ms. "Sur les oiseaux," and a group of holograph notes (10 leaves, 33 p.); in French.
MSS 584 A

607. Puccinotti, Francesco, 1794-1872. *Papers.* 1821-1822. 3 items.

Latin document (1822 April [1?]) concerning Puccinotti, note (partly illegible, undated), and "Elenco de requisiti originali e legalizzati appartenenti al Dottore Francesco Puccinotti" (1821 Oct. 20) by Sabbatini; in Italian.
MSS 1178 A

608. Underwood, Thomas Richard, 1772-1835. *Letter.* 1821. 1 item (2 p.)

A.L.S. (1821 June 17 London) to Ampère concerning a conversation with Davy regarding magnetic electricity and the Voltaic pile; in French.
MSS 1487 A

609. Goethe, Johann Wolfgang von, 1749-1832. *Paper.* 1822. 1 item (2 p.)

Holograph leaf from sometime between July and Dec., 1822; in German.
MSS 603 A

610. De La Beche, Henry Thomas, Sir, 1796-1855. *Letter.* 1822. 1 item (1 p.)

A.L.S. (1822 June 20, Bristol) to an unnamed correspondent regarding payment for the printing of plates.
MSS 418 A

611. *List of double stars &c observed with a 3 1/2 inch achromatic telescope of Professor Barlow's construction.* [after 1822]. 1 item (4 p.)

Manuscript copy of Barlow's observations. Formerly laid in Basil Hall's *Magnetical Observations.*
MSS 912 A

612. Moll, Karl Ehrenbert, Ritter von, 1760-1838. *Letter.* 1822. 1 item (2 p.)

A.L.S. (1822 Mar. 4) to Hausmann mentioning Stromeyer; in German.
MSS 1021 A

613. Poncelet, Jean Victor, 1788-1867. *Letters.* 1822-1852. 12 items.

Twelve A.L.S., 10 of them (1822-1831, Metz) to General Baudraud, one (1852 June 17) to Guerin and one (undated) to a confrère with a biographical note on Poncelet added (1862 Mar. 12, Florence) by I.I.S.(?); in French.
MSS 1158 A

614. Ampère, André Marie, 1775-1836. *Cours d'analyse et de mécanique à l'école polytechnique [par] M. Ampère, professeur, 1re année d'études, 1823-1824.* 1823-1824. [2], 444 p.: ill.

Mathematical diagrams throughout. Table of contents, 1 folded sheet (4 p.) laid in. Title page signed "G. Vincens."
MSS 8 B

615. Bolyai, Farkas, 1775-1856. *Correspondence.* 1823-1832. 2 items.

Transcript of A.L.S. (1832 Mar. 6) from Gauss to Bolyai; facsimile of letter from J. Bolyai to F. Bolyai (1823 Nov. 3); in Hungarian. Additional material in German and French.
MSS 137 A

616. Brandes, Rudolf, 1795-1842. *Letters.* 1823-1832. 2 items.

Two A.L.S. (1823 April 28 and 1832 April 2, Salzuflen), one to Döbereiner; in German.
MSS 170 A

617. Kastner, Karl Wilhelm Gottlob, 1783-1857. *Letters.* 1823-1836. 5 items.

Five A.L.S. to various correspondents including Dulk; in German.
MSS 772 A

618. Libri, Guillaume, 1803-1869. *Letters.* [between 1823 and 1869]. 3 items.

Three A.L.S., one dated 1858 May 6, London, and one to Breschet; in French.
MSS 892 A

619.　Richardson, Thomas, 1771-1853. *Letter.* 1823. 1 item (2 p.)

A.L.S. (1823 Apr. 23, Stamford Hill) to Samuel Thoroughgood inviting him to meet and befriend Stephenson.　　　　　　　　　　　MSS 1211 A

620.　Snell, Friedrich Wilhelm Daniel, 1761-1827. *Letter.* 1823. 1 item (1 leaf)

A.L.S. (1823 May 27, Giessen) to André mentioning André's visit to Stuttgart; in German.　　MSS 1386 A

621.　Everett, Edward, 1794-1865. *Document.* [between 1824 and 1834]. 1 item (1 p.)

D.S. inviting Dixwell to the Cambridge, Mass., Observatory.　　　　　　　　　　　MSS 493 A

622.　Jacobi, Karl Gustav Jakob, 1804-1851. *Papers.* [between 1824 and 1851]. 4 items.

A.L.S. (1847 Nov. 22, Berlin) to Dr. Barthold, 2 leaves of autograph notes and calculations, and an autograph signature.　　　　　　　　　　　MSS 743 A

623.　Lenz, Heinrich Friedrich Emil, 1804-1865. *Letter.* [between 1824 and 1865]. 1 item (1 leaf)

A.L.S. (undated); in German.　　MSS 881 A

624.　Mollweide, Karl Brandon, 1774-1825. *Letter.* 1824. 1 item (1 leaf)

A.L.S. (1824 Feb. 20, Leipzig); in German.　　　　　　　　　　　MSS 1022 A

625.　*Notes on chemistry &c.* [between 1824 and 1855?]. 385 [i.e., 393] p. [i.e., 197 leaves] 25 cm.

Topics include matter, galvanism, electricity, organic chemistry, and the circulation of the blood. Discussions of individual elements include experiments. The list of discoveries of metals (p. 165) ends with 1824. The section on aluminum (p. 229) cites Davy's discovery as "all that is known of aluminum." Spine title.　　MSS 1094 B

626.　Ramognini, Antonio. *Institutiones physicae generalis et particularis auctore Antonio Ramognini, C.M.P.; adiecto in fine Tractatu de electricitate Josephi Scarzolo, C.M.P.* 1824. 510 p.: chiefly col. ill., maps; 22 cm.

Manuscript on physics and electricity with a third text, "Tractatus de astronomie"; Latin text with extensive Italian quotations. Title statement concludes: quae omnia pubblice(!) ab iisdem auctoribus tradebantur Savonae in N. Congregationis Missionis Collegio Anno 1824.　　　　　　　　　　　MSS 1261 B

627.　Stevenson, Robert, 1772-1850. *Papers.* [between 1824 and 1850]. 3 items.

L.S. with envelope (1841 Jan. 23, Edinburgh) to Messrs. Langton and Bicknell soliciting a bid for the sale of spermaceti oil, on the letterhead of "Northern Lights" with the seal of "Northern Lighthouses, Engineers Office," and an anonymous ms. ("Balance Crane") describing plate XVII in Stevenson's *An Account of the Bell Rock Lighthouse* (1824).　　MSS 1418 A

628.　Warburton, Henry, 1784?-1858. *Letter.* 1824. 1 item (2 p.)

A.L.S. (1824 Jan. 11) to Pond at the Royal Observatory at Greenwich concerning the acids used in engraving.　　　　　　　　　　　MSS 1531 A

629.　Franklin, John, Sir, 1786-1847. *Papers.* 1825-1834. 5 items.

Two A.L.S., 2 D.S. accounting for wages and materials, and autograph signature.　　　MSS 542 A

630.　Lax, William, 1761-1836. *Letter.* 1825. 1 item (3 p.)

A.L.S. (1825 Nov. 20) to Mrs. Pond concerning Lax's recent visit.　　　　　　　　　MSS 861 A

631.　Young, Thomas, 1773-1829. *Letter.* 1825. 1 item (1 leaf)

A.L.S. (1825 Nov. 8, London) to Professor Schumacher concerning astronomical computations and instruments.　　　　　　　　　MSS 1592 A

632.　Aikin, Arthur, 1773-1854. *Admission ticket to the distribution of the rewards of the Society of Arts. . .* 1826. 1 item (1 p.): col. ill.

Printed ticket signed (1826 May 29, London).

　　　　　　　　　　　MSS 70 A

633. Brongniart, Adolphe Théodore, 1801-1876. *Letters.* [between 1826 and 1876]. 3 items.

Three A.L.S. (one dated 1835 Feb. 3) to various correspondents including Duméril; in French.　　MSS 186 A

634. Brugnatelli, Gaspare, 1795-1852. *Letter.* 1826. 1 item (1 p.)

A.L.S. (1826 July 18, Pavia) to the Società Tipografica di Classici Italiani, Milan; in Italian.　　MSS 189 A

635. Buchner, Johann Andreas, 1783-1852. *Letter.* 1826. 1 item (1 p.)

A.L.S. (1826 July 16) to Weisbrunner(?); in German.　　MSS 197 A

636. Bull, Ephraim Wales, 1806-1895. *Card.* [between 1826 and 1895]. 1 item (1 p.): col. port.

One card with autograph signature. Additional material includes newspaper clipping.　　MSS 202 A

637. Gillet de Laumont, François Pierre Nicolas, 1747-1834. *Letter.* 1826. 1 item (3 p.)

A.L.S. (1826 Feb. 16, Paris) to Payen concerning experiments on the effect of muriated limestone and chloride of lime on soil fertility; in French.　　MSS 854 A

638. Jameson, Robert, 1774-1854. *Letters.* 1826-1833. 2 items.

A.L.S. (1826 Oct.) to Mylne (i.e., Home) on comets, and A.L.S. (1833 June 29, Edinburgh) to Clift concerning a fossil bone, with Clift's response.　　MSS 749 A

639. Lenoir, Etienne, 1744-1832. *Letter.* 1826. 1 item (3 p.)

A.L.S. (1826 Aug. 19, Paris) to Gillet de Laumont concerning galvanizing following Davy's process; in French.　　MSS 880 A

640. South, James, Sir, b. 1785. *Letters.* 1826-1877. 12 items.

Twelve A.L.S. (some from the Observatory, Kensington) to various correspondents including Prof. Moll, W. Basting (about Faraday), the Assistant Secretary of the Royal Astronomical Society, and Sharpey.　　MSS 1393 A

641. Stephenson, Robert, 1803-1859. *Letters.* 1826-1856. 7 items.

Six A.L.S. to Thomas Coates, Henry Cole, Edward Lally(?), John Lucas, Allen Ransome, and Stephenson's father George Stephenson, and holograph pass S. on letterhead of the London & Birmingham Railway.　　MSS 1415 A

642. Airy, George Biddell, Sir, 1801-1892. *Letters.* 1827-1883. 69 items.

Sixty-three A.L.S. to various recipients, including Arago, Beaufort, Hogg, Lubbock, and Ørsted; 2 D.S. acknowledging receipt of books for the Royal Observatory; holograph copy of letter to Comptroller General of the Exchequer; and 3 envelopes.　　MSS 61 A

643. Erdmann, Otto Linné, 1804-1869. *Letter.* 1827. 1 item (1 p.)

A.L.S. (1827 June 4, Dresden); in German.　　MSS 486 A

644. Francoeur, Louis Benjamin, 1773-1849. *Letters.* 1827-1833. 3 items.

Three A.L.S. (two dated 1827 Dec. 5 and 1833 Mar. 10) to de Savallette, Vallot, and Puissant; in French.　　MSS 535 A

645. Horner, Johann Casper, 1774-1834. *Letter.* 1827. 1 item (4 p.)

A.L.S. (1827 Sept. 5, Zurich) to Tilesius describing Horner's private and professional life and recent trip to Geneva; in German.　　MSS 722 A

646. Orbegozo, J. de. *Papers on astronomy.* 1827-1828. 2 items; 26 cm.

A.L.S. (1828 Sept. 6, Orizaba) to [Manuel?] Teran on formulas for calculations concerning the sun and moon. With "Tabla dela paralaje dela luna [q--1?] comprenden deide los 45° haitalos 65° de altura aparente y citan llevadas haita dos decimales desiginidos por J. O. en 1827"; in Spanish. Cover title: Cartas del [g--1?] Orbegoso.　　MSS 1102 B

647. Pfaff, Johann Wilhelm Andreas, 1774-1835. *Papers.* 1827-1830. 3 items.

Two A.L.S. (1827, 1828) and A. ms. S. (1830 April 13): [Vorrede zu] W. Herschels Schriften II Band, das Wesen der Sterne und der Georg Stern; all written at Erlangen; in German.   MSS 1132 A

648. Sertürner, Friedrich Wilhelm, 1783-1841. *Letter.* 1827. 1 item (3 p.)

A.L.S. (1827 Feb. 12, Hameln) mentioning Germany, France, and England; in German.   MSS 1356 A

649. Trommsdorff, Johann Bartholomäus, 1770-1837. *Letter.* 1827. 1 item (1 leaf)

A.L.S. (1827 Nov. 22, Erfurt) to Döbereiner; in German.   MSS 1483 A

650. Vigors, Nicholas Aylward, 1785-1840. *Letter.* 1827. 1 item (1 leaf)

Form letter completed and signed (1827 Nov. 5, Zoological Society) to Charles Parker acknowledging the gift of "a living specimen of a Pecari" to the Menagerie.   MSS 1508 A

651. *Zeichnungen und Notizen von den Arbeiten an dem Gange unter der Thems.* 1827. [14] p.: ill. (one col.); 13 x 16 cm.

Bound ms. (1827 Sept.) concerning work on the Thames Tunnel from 1799 through 1827; includes 4 ink and wash illustrations; in German. Title on box: Gange unter der Thems.   MSS 1293 B

652. Encke, Johann Franz, 1791-1865. *Papers.* 1828-1856. 12 items.

Holograph article, "Ueber Herrn von Zachs astronomische Statigkeit" (1832), 8 A.L.S., 2 holograph pages, and 1 calling card; in German.   MSS 482 A

653. Gruithuisen, Franz von Paula, 1774-1852. *Papers.* 1828-1840. 6 items.

Four A.L.S., and 2 mss.; in German.   MSS 628 A

654. Hamilton, William John, 1805-1867. *Envelope.* 1828. 1 item (1 p.)

Portion of envelope (1828 May 17) addressed to Mrs. Hamilton, Lama Place, London.   MSS 659 A

655. Jacobi, Karl Friedrich Andreas, 1795-1855. *Letter.* 1828. 1 item (2 p.)

A.L.S. (1828 Feb. 6, Landesschule Pforta) to Bibliothekssekretär Möller, Gotha; in German.   MSS 742 A

656. Reich, Ferdinand, 1799-1882. *Letters.* 1828-1829. 4 items.

Four A.L.S. (1828-1829, Freiberg) concerning journals and books, especially Sowerby's *Mineral Conchology of Great Britain;* in German.   MSS 1197 A

657. Sabine, Edward, Sir, 1788-1883. *Letters.* 1828-1854. 9 items.

One A.L.S. (1829 Dec. 27) to Arago concerning the Aurora Borealis and magnetism, in French, 1 printed form letter of the Royal Society (1828 June 13) signed by Sabine as secretary, to Rev. Josiah Forshall, 5 A.L.S. to various correspondents including J.B. Yates, Dr. Neumeyer, Gassiot, and perhaps Lubbock, and 1 A.L.S. (1838 Feb. 20) enclosing 1 A. ms. S. (3 p.) concerning magnetic observations in India.   MSS 1314 A

658. Traill, Thomas Stewart, 1781-1862. *Letter.* 1828. 1 item (2 p.)

A.L.S. (1828) to William Rathbone concerning Traill's inability to find purchasers for Audubon's paintings.   MSS 1479 A

659. Babbage, Charles, 1792-1871. *Papers.* 1829-1863. 13 items.

Various A.L.S. including one of 9 pages (1835 Aug. 2) concerning modifications to the calculating machine. Finding aids, obituary, biography, and portrait in folder.   MSS 32 A

660. Brunel, Isambard Kingdom, 1806-1859. *Letters.* 1829-1855. 9 items.

Nine A.L.S. to various correspondents including E. Clerk(?).   MSS 190 A

661. Cobbett, William. *Letter.* 1829. 1 item (7 p.)

A.L.S. (1829 Nov. 11, London) to Crawshay concerning his father's studies on Indian corn (zea) in England.   MSS 356 A

662. Daguerre, Louis Jacques Mandé, 1789-1851. *Letter.* 1829. 1 item (1 p.)

A.L.S. (1829 Sept. 22, Paris) to M. Dumersan; in French.　　　　MSS 392 A

663. Despretz, César Mansuète, 1789-1863. *Addenda to Résume des travaux de physique.* [between 1829 and 1863]. 38 p.; 27 cm.

Copy of a summary (31 p.) with holograph addenda (pp. 32-8 and tipped in at p. 11, 13, and 14) including an A.L.S. to Tyndall written on a transcript of a letter to Grove, undated; in French.　　　　MSS 254 B

664. Eaton, Amos, 1776-1842. *Letter.* 1829. 1 item (3 p.)

A.L.S. (1829 Dec. 15) to Mr. Little concerning his subscription to the *New York Journal.*　　　　MSS 467 A

665. Fechner, Gustav Theodor, 1801-1887. *Papers.* 1829-1864. 6 items.

Two A.L.S., 3 A. notes, and 1 holograph (10 p.); in German.　　　　MSS 504 A

666. Geoffroy Saint Hilaire, Isidore, 1805-1861. *Papers.* 1829-1858. 17 items.

Sixteen A.L.S. to various correspondents, and a holograph sheet (draft of letter?); in French.　　　　MSS 585 A

667. Gerstner, Franz Anton, Ritter von, 1795-1840. *Papers.* 1829-1831. 4 items.

Three A.L.S. to various correspondents including Cordier, and a draft of a review of his father's *Handbuch der Mechanik;* in German.　　　　MSS 590 A

668. Sartorius Von Waltershausen, Wolfgang, 1809-1876. *Letter.* [between 1829 and 1876]. 1 item (2 p.)

A.L.S. to Legations Rath Kestner(?); in German.　　　　MSS 1326 A

669. Senefelder, Alois, 1771-1834. *Letter.* 1829. 1 item (2 p.)

A.L.S. (1829 Oct. 30, Munich) to Sachse(?); in German.　　　　MSS 1355 A

670. Sömmering, Samuel Thomas von, 1755-1830. *Letter.* 1829. 1 item (2 p.)

A.L.S. (1829 Mar. 24, Munich); in German.　　　　MSS 1388 A

671. Becquerel, Antoine César, 1788-1878. *Letters.* 1830-1877. 6 items; port.

Five A.L.S. including 2 to Milne-Edwards, explaining Becquerel's sudden departure from Paris, and requesting a subsidy for his research assistant's work on capillary tubes, and autograph; in French.　　　　MSS 60 A

672. Cuvier, Frédéric Georges, 1773-1838. *Letters.* 1830-1845. 2 items.

Two A.L.S. (1830 June 9 and 1845 March 15), the former on letterhead of the autograph collection of Max Thorell(?), the latter making an appointment with Sauvageot; in French.　　　　MSS 389 A

673. Hansen, Peter Andreas, 1795-1874. *Letters.* 1830-1867. 8 items.

Eight A.L.S. including one to Weber; in German.　　　　MSS 660 A

674. Henry, William Charles, 1804-1892. *Letter.* [1837?]. 1 item (2 p.)

A.L.S. ([1837?] Aug. 14, Manchester) to Professor Hofer introducing Liebig.　　　　MSS 688 A

675. Nürnberger, Joseph Christian Emil, 1779-1848. *Letters.* 1830-1847. 3 items.

Two A.L.S. (1830 Nov. 6, and 1847 May 1) and A. paper (undated); in German.　　　　MSS 1081 A

676. Peirce, Benjamin, 1809-1880. *Letter.* 1830-1880. 1 item (1 leaf)

A.L.S. (n.d.) to R. P. Ames, a man whose mother he praises.　　　　MSS 1107 A

677. Phillips, John, 1800-1874. *Letters.* 1830-1842. 4 items.

One signature and 3 A.L.S., one (1830 Nov. 22, Yorkshire Museum) to Sowerby, one (1842 [May?] 15, Malvern) to Rev. Dr. Buckland mentioning the work of Strickland, and one (April 24) to R. H. Collins arranging a visit by "His R. H.," possibly the prince consort.

MSS 1133 A

678. Powell, Baden, 1796-1860. *Letters.*
1830-1850. 3 items: ill.

A.L.S. (1830 March 3, Oxford) to Gilbert recommending Mr. Griswill for membership in the Royal Society, A.L.S. (1848 Feb. 6, Oxford) to an unnamed correspondent concerning meteors with a diagram, and A.L.S. (1850 May 29, London) to an unnamed correspondent concerning meteors.　　　　　　　　MSS 1167 A

679. Regnault, Henri Victor, 1810-1878.
Letters. [between 1830 and 1878]. 6 items.

A.L.S. (1859 May 14) to "Monsieur l'abbé," 1 A.L.S. (186-?) to a general on letterhead of the Manufacture Impériale de Porcelain, 1 A.L.S. (184-?) to Varcollier on letterhead of the Ecole Royale Polytechnique recommending Bourson, and 2 A.L.S. (undated, one with envelope) to de Sénarmont requesting chemicals and to Boutroy-Charlard on the instrument maker Fastre; in French.
　　　　　　　　MSS 1196 A

680. Ressel, Josef, 1793-1857. *Correspondence.* 1830-1863. 11 items.

Three A.L.S. (1834-1843) to Johann Tischler, two A.L.S. (1830-1860) to Franz Skola, two A.L.S. (1861-1863) from Skola to Ressel, one A. leaf of notes, 2 A.L.S. (1861-1863) from Cervio Reggio to Skola on letterhead of the Comité für das Ressel-Monument, and A. envelope with seal of the Comité to Skola; in German.
　　　　　　　　MSS 1206 A

681. Rüppell, Eduard, 1794-1884. *Letters.*
1830-1836. 4 items.

Four A.L.S. to F. G. Levrault concerning books; in French.　　　　　　　　MSS 1244 A

682. Silliman, Benjamin, 1779-1864. *Letters.*
1830-1856. 6 items: ill.

Six A.L.S., all from New Haven, including one to Wightman concerning electro-magnetic apparatus and one to Eaton containing a sketch.　　MSS 1373 A

683. Somerset, Edward Adolphus Seymour, 11th duke of, 1775-1855. *Letter.* 1830. 1 item (1 leaf)

A.L.S. (1830 Dec. 26, Park Lane, [London]) to Faraday thanking him for his "new edition" and praising its introduction and "the mode of reference which pervades the whole."　　　　　　　　MSS 1389 A

684. Wheatstone, Charles, Sir, 1802-1875.
Papers. 1830-1879. 22 items: ill., port.

Twenty-one A.L.S. (1830-1875), one in French, to various correspondents including Frankland (1860), Home (1843), Lubbock (1845, 1851), and Moigno (1852), and D.S. (1879(!) July 28, Paris) concerning the Institut de France, Académie des sciences. Topics include photography, electro-magnetism, telegraphy, scientific apparatus, the Atmospheric Railway, and the stereoscope. Additional material includes 2 photographic portraits; one perhaps is of Wheatstone.　　　　MSS 1558 A

685. Celestin, Galli. *Letter.* 1831. 1 item (8 p.)

A.L.S. (1831 June 21, London) to the Duke of Sussex concerning Celestin's stenography machine; in French.
　　　　　　　　MSS 319 A

686. Faraday, Michael, 1791-1867. *Letters.*
1831-1835. 2 items.

Two A.L.S., each with seal, to Phillips. The first (1831 Nov. 29, Brighton) outlines his success in producing electricity from magnetism and gives title and contents of the book he wrote immediately afterwards: *Experimental researches in electricity.* The second (1835 Ju(sic) 12, London) is a note: "Are the enclosed all right"(sic). Additional material: information on electromagnetic induction, and xeroxed copies of photographs of the letters. [XII]
　　　　　　　　MSS 853 B

687. Julia de Fontenelle, Jean Sébastien Eugène, 1790-1842. *Papers.* 1831. 2 items.

A.L.S. (1831 Jan. 21, Paris) offering his services as lecturer on applied chemistry at the Association Polytechnique, and holograph ms. "Mémoire Chimico Médical et Pharmaceutique sur la Moutarde" (4 p.), published in the *Journal de Chimie Médicale;* in French.　　MSS 762 A

688. Matteucci, Carlo, 1811-1868. *Letters.* [between 1831 and 1868]. 5 items.

Five A.L.S., one dated 1833 April 10, one to Gazzeri; one in French, and four in Italian.   MSS 975 A

689. Orfila, Matthieu Joseph Bonaventure, 1787-1853. *Letters.* 1831-1847. 12 items.

Twelve A.L.S. to various correspondents including Barbé-Marbois, David d'Angers, and Trémont; in French.   MSS 1087 A

690. Schomburgk, Moritz Richard, 1811-1891. *Papers.* [between 1831 and 1891]. 3 items.

A.L.S. ([18]67 Feb. 26, Botanic Gardens, Adelaide), printed L.S. (Bontanischer Garten, Adelaide, Sued Australien, i.e., South Australia), both in German, and holograph ms. abstract of his paper "On the Superstitions and the Astronomical Knowledge of the Indians of Guiana" (3 p.) concluding with an A. note to Mr. Booth.   MSS 1336 A

691. Somervile, Mary, 1780-1872. *Letters.* 1831-1840. 6 items.

Six A.L.S. to various correspondents including Monsignore de Medici Spada and Mary Horner.   MSS 1390 A

692. Whewell, William, 1794-1866. *Letters.* 1831-1861. 5 items.

Five A.L.S. to Hodgeson ([18]31), Murchison (1841), Lubbock (1846, 1847), and Richard Bentley (1861).   MSS 1559 A

693. Bache, Alexander Dallas, 1806-1867. *Letters.* 1837-1858. 5 items.

A.L.S. (1837 March 10, Washington) from Bache to Hon. H. A. J. Dearborn; A.L.S. ([18--?] Dec. 6) to unidentified introducing Charles Schott; A.L.S. (1858 March 13) to J. W. Cochran stating his satisfaction with Cochran's gauging instrument; A.L.S. (1858 March 15) marked "Copy" to Hon. Howell Cobb, Secretary of Treasury, recommending Cochran's instrument; A.L.S. (1857 March 16) to A. Hanson Foote concerning the 1855 Coast Survey Report. Photocopy of letter to Foote in Folder.   MSS 34 A

694. Binney, Edward William, 1812-1881. *Letter.* [between 1852 and 1881]. 1 item (1 p.)

A.L.S. (undated) to Mrs. Paget accepting a dinner invitation.   MSS 115 A

695. Brown, Robert, 1773-1858. *Letters.* 1832-1857. 3 items.

Three A.L.S., one (1857 Aug. 10, London) to Humboldt introducing Mr. Polyblank, a photographer.   MSS 188 A

696. Butler, Eleanor, 1745?-1829. *Letter.* [1832?]. 1 item (2 p.)

A.L.S. (18[32?] Nov. 19) to Lady Dungannon mentioning General and Lady Elizabeth Fitz Roy and their grief. Signed: E. Butler & S. Ponsonby. On paper with watermark date 1823. Pencilled at bottom: Plas Nwyd Langollen [i.e., Plasnewydd cottage, Langollen, Wales].   MSS 1607 A

697. Dulong, Pierre Louis, 1785-1838. *Letters.* 1832-1839. 4 items.

Four A.L.S. to Carus, Dumas, Ørsted, and Hericart-Ferrand; in French.   MSS 462 A

698. Gray, Asa, 1810-1888. *Papers.* 1832-1874. 10 items.

Eight A.L.S. to various correspondents including Peabody and Williamson, 1 autograph signature, and a holograph leaf S. concerning the watering of plants.   MSS 614 A

699. Henry, Joseph, 1797-1878. *Correspondence.* 1832-1877. 35 items.

Nineteen A.L.S., 13 L.S., and 3 D.S. to various recipients including Cleaveland, Earle, Greble, and Rogers. Photocopies in folder 2.   MSS 686 A

700. Lubbock, John William, 1803-1865. *Letters.* 1832-1860. 9 items.

Six A.L.S. to various correspondents including Herschel and Kingsley, 2 zincographs concerning a Cambridge University election, and an autograph fragment with signature to the executors of Chantrey.   MSS 928 A

701. *Meccanica.* 1832. 1231 p.: ill.; 30 cm.

Manuscript on mechanics, dynamics, statics, and motion, with an index of "Proposizioni di meccanica." The order of the two sections of the text is apparently transposed with "Sezione prima, statica" following "Dinamica, sezione IIa." The section on dynamics ends with the colophon "Fine del Meccanica" and the section on statics has 2 colophons, dated July and December 1832. Spine title: Meccanica / Sublime / 3°Corson / [S---?] Cremona.   MSS 1260 B

702. Morse, Samuel Finley Breese, 1791-1872. *Papers.* 1832-1871. 22 items: port.

Sixteen A.L.S. (one in Italian) to various correspondents including Arago, Earle, Hart, Hubbell, Rosetti, and Sprague, 3 A.S., D.S., A. label S., and a carte de visite photograph. Additional material in second folder.   MSS 1042 A

703. Foerstemann, Wilhelm August, 1791-1836. *Letter.* 1832. 1 item (2 p.)

A.L.S. (1832 June 18, Danzig) to the editor of *Jenaische Literaturzeitung* concerning Foerstemann's book on the antagonism of positive and negatuve numbers; in German.   MSS 522 A

704. Sedgwick, Adam, 1785-1873. *Letters.* 1832-1871. 7 items.

Six A.L.S. to various correspondents including Rev. William Robinson, Sowerby, and Webster, and autograph signature (1870).   MSS 1348 A

705. Silliman, Benjamin, 1816-1885. *Papers.* 1825-[1974?]. One box; 33 cm.

Several hundred items, mostly correspondence, at least 3 mss. (1 in German), various notes, a telegram, and a document. Much of the correspondence is with relatives and descendants, including a Frenchman (2 A.L.S., 1881, in French). 2 A.L.S. from W. Feddersen (1881, Leipzig); in German. Collection includes correspondence of both father and son.   MSS 1280 B

706. Stephenson, George, 1781-1848. *Letters.* 1832-1847. 3 items.

A.L.S. (1832 Jan. 25, Liverpool) to Philip Madison, A.L.S. (1836 May 13) to Robert Stephenson, and A.L.S. (1847 June 10, Taptonhouse) to J. and [C.?] Davies.   MSS 1414 A

707. Wilberforce, William, 1759-1833. *Letter.* 1832. 1 item (4 p.)

A.L.S. (1832 Mar. 18, St. Boniface near Newport, Isle of Wight) to [J. J. Buxton?] concerning family matters.   MSS 1570 A

708. Agassiz, Louis, 1807-1873. *Letters.* 1833-1873. 27 items: ports.

Twenty-three A.L.S. (some in German and French); 2 signed memoranda; 2 D.S.; and 2 envelopes.   MSS 69 A

709. Boguslawski, Palon Heinrich Ludwig von, 1789-1851. *Papers.* 1833-1850. 5 items.

Four A.L.S. to various correspondents including Eichstädt and the *Allgemeine Anzeiger oder National Zeitung, Gotha* and a holograph ms. (1833 Dec.) entitled: "Noch ein Blick auf unsere deutschen Universitäten"; in German.   MSS 133 A

710. Jackson, Richard. *Memorandum of Agreement [concerning specifications for a locomotive steam engine].* 1833. 1 item (3 p.): ill.; 43 cm.

D.S. Memorandum of Agreement from Fenton, Murray & Jackson, engineers of Leeds, to William C. Molyneux, merchant of Liverpool, concerning the specifications for a locomotive steam engine. Written on 1833 Aug. 1, the document was signed by Jackson on 1833 Aug. 3. Also added then to the same page is a letter to Molyneux, referring to the diagram, etc., and signed by J. M. Dickinson.   MSS 852 B

711. Lindley, John, 1799-1865. *Document.* 1833. 1 item (1 leaf)

A.D.S. (1833 Feb. 2, Chiswick) to J. B. Nichols concerning the Horticultural Society's garden shipment of apple and pear grafts.   MSS 900 A

712. Marchand, Richard Félix, 1813-1850. *Papers.* [between 1833 and 1850]. 3 items.

Two A.L.S. (one dated 1847 Feb. 10, Halle), and 1 holograph ms. (4 p.); in German.   MSS 957 A

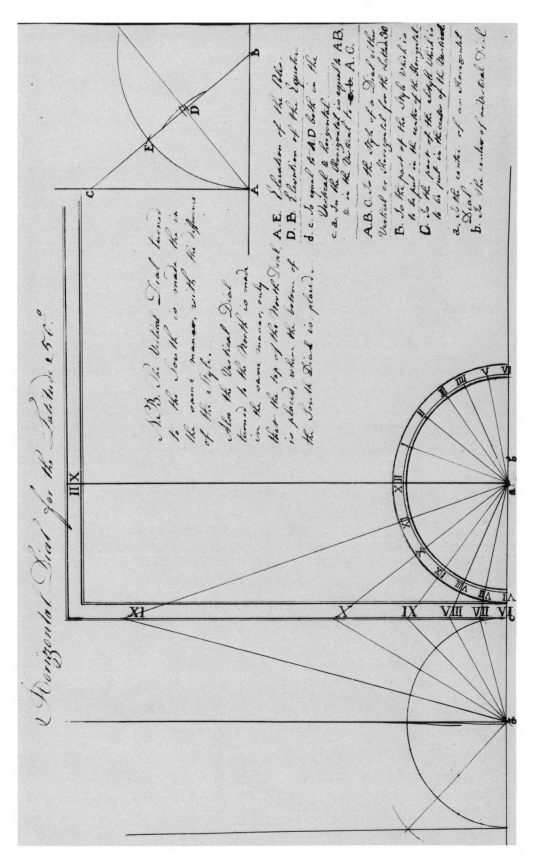

VIII A.   Drawing of a sundial (18th century). Horizontal Dial for the
Latitude 50.
Entry No. 224, SI Neg. 85-9513.

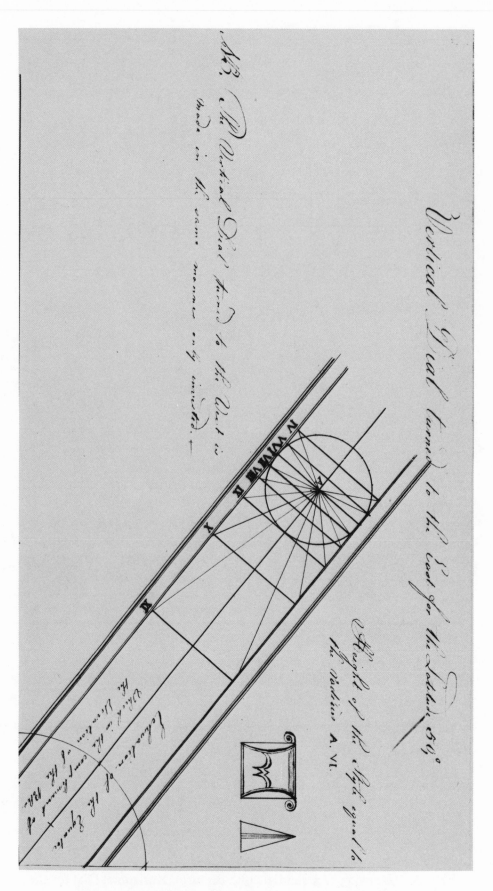

B. Drawing of a sundial (18th century). Vertical dial turned to the
    East for latitude 50.
    Entry No. 224, SI Neg. 84-7176.

713. Payen, Anselme, 1795-1871. *Letter.* 1833. 1 item (2 p.)

A.L.S. (1833 Aug. 8, Paris) transmitting a report of Dumas concerning vegetable physiology; in French.

MSS 1105 A

714. Pouillet, Claude Servais Mathias, 1790-1868. *Letters.* 1833-1859. 3 items.

A.L.S. (1833 Dec. 17) to Gaultier de Claubry praising his research and asking him to assist Melloni, A.L.S. ([18]39 Dec. 4) to Cordier, both on letterhead of the Conservatoire Royal des Arts et Metiers, and A.L.S. ([18]59 Oct. 7, Epinay) to Sénarmont, President de l'Institut; in French.

MSS 1163 A

715. Reichenbach, Karl Ludwig Friedrich, Freiherr von, 1788-1869. *Letters.* 1833-1835. 3 items.

Three A.L.S. (1833 Jan. 24, Blansko, [18]35 Nov. 19, Reisenberg, and undated) to Herr [Burkhandler?] Heubner; in German.

MSS 1200 A

716. Rennie, George Banks, 1791-1866. *Letters.* 1833-1850. 4 items.

A.L.S. (1833 Oct. 11, Whitehall Place) to Charles Lawrence, A.L.S. (1837 Aug. 18, London) to Vogel de Vogelstein, A.L.S. (1848 Dec. 5, Whitehall Place) to James Lich concerning electric light, and A.L.S. (1850 May 24, Holland Street) paying a bill.

MSS 1203 A

717. Rennie, John, Sir, 1794-1874. *Letter.* 1833. 1 item (1 leaf)

A.L.S. (1833 June 11, Holland Street, [London]) to Robertson about railways and canals.

MSS 1205 A

718. Samuda, Joseph d'Aquilar, 1813-1885. *Letter.* [between 1833 and 1844?]. 1 item (1 leaf)

L. signed "Samuda Bros." to the editor of the *Railway Magazine,* replying to an anonymous letter printed in a recent issue of the magazine; pasted to scrapbook page. With 2 apparently unrelated ms. leaves pasted to verso of scrapbook page.

MSS 1324 A

719. Baily, Francis, 1774-1844. *Letter.* 1834. 1 item (1 p.)

A.L.S. (1834 July 25) to Rev. R. Sheepshanks declining an invitation to view a "large equatorial" at Camden Hill unless he is invited by both parties.

MSS 38 A

720. Beaufoy, Mark, 1764-1827. *[Order book for orders received for Beaufoy's Nautical Experiments,* vol. 1]. 1834-1835. [32] leaves, 32 cm.

Two sheets of notes laid in.

MSS 263 B

721. Biela, Wilhelm von, 1782-1856. *Letters.* 1834-1843. 3 items.

Two A.L.S. (1834 May 24; 1843 Nov. 15) with envelope; in German.

MSS 112 A

722. Buch, Leopold von, Freiherr, 1774-1853. *Letter.* 1834. 1 item (1 p.): port.

A.L.S. (1834 May 6) to Baer; in German.

MSS 195 A

723. Buckland, William, 1784-1856. *Letters.* 1834-1849. 6 items.

Five A.L.S. to various correspondents including Ranson and Crosse, and an envelope. Additional material includes a newspaper obituary.

MSS 200 A

724. Edward, Thomas, 1814-1886. *Autograph signature.* [between 1834 and 1886]. 1 item (1 p.): port.

Newspaper obituary included.

MSS 471 A

725. Fuchs, Johann Nepomuk von, 1774-1856. *Letter.* 1834. 1 item (3 p.)

A.L.S. (1834 Nov. 21, Munich); in German.

MSS 555 A

726. Harrison, Thomas Elliott, 1808-1888. *Letter.* 1834. 1 item (2 p.)

A.L.S. (1834 Jan. 18, Maidstone) to Pearson recommending Sturgeon as a lecturer on electricity and galvanism.

MSS 1407 A

727. Kuhn, Karl Amandus, d. 1848. *Bergbaukunst.* 1834-1835. 474 p.: ill.; 23 cm.

Manuscript treatise on the science of mining. Wide margins throughout with illustrations, especially of tools, and some citations, dated 1817-1818. In pencil, a later hand numbered the rectos 1-467 [i.e., 471] and added 31 section divisions in roman numerals. On second leaf, in ink, is the name Sigmund Geiges and title, Bergbaukunst I, attributed to Prof. Kuhn, 1834-1835; in German.

MSS 836 B

728. Milne-Edwards, Henri, 1800-1885. *Letter.* 1834. 1 item (3 p.)

A.L.S. (1834 Dec. 2, Versailles) to Arcet; in French.

MSS 1012 A

729. Purkyné, Jan Evangelista, 1787-1869. *Letters.* [1834]-1858. 2 items.

A.L.S. (1858 March 23, Prague) and A.L.S. (undated, text mentions 1834) to Otto; in German.   MSS 1182 A

730. Servan, abbé. *Letter.* 1834. 1 item (2 p.)

A.L. signed "de Servan" (1834 Feb. 4) to an unnamed correspondent concerning meeting for a cup of coffee; in French.   MSS 1357 A

731. Struve, Friedrich Georg Wilhelm, 1793-1864. *Correspondence.* 1834-1851. 4 items.

Three A.L.S. (1834-1836) to T. v. Ertel of the "mathematischen Instituts," and A.L.S. (1851 Dec. 19, Pulkovaden) to an unnamed correspondent whose reply is on the same sheet; in German.   MSS 1431 A

732. Watt, James, 1769-1848. *Letters.* 1834-1847. 6 items.

One A.L.S. (1834 Apr. 1) to Ebenezer Robins concerning some plans, one A.L.S. (1834 Apr. 16) and one L.S. (1847 Sept. 7) to the Canal Office of Birmingham acknowledging the receipt of dividends, and two A.L.S. (1839 Oct. 21 and 31) and one L.S. (1846 Oct. 10) to J. W. Croker concerning Watt's famous father and a book ready for publication.   MSS 1538 A

733. Barlow, Peter, 1776-1862. *Letters.* 1835. 2 items.

Two A.L.S. including one (1835 Jan. 14) to Stark.

MSS 48 A

734. Blake, James, 1815-1893. *Letters.* [between 1835 and 1893]. 2 items.

Two A.L.S. (April 25 and Sept. 18, London) to Pelouze; in French.   MSS 119 A

735. Brandes, Karl Wilhelm Hermann, 1814-1843. *De litterarum [maino?] et impressis matheseos studio non spernendo.* 1835. 1 item (15 p.)

Holograph signed concerning the importance of mathematical studies; in Latin.   MSS 167 A

736. Configliachi, Luigi, 1787-1864. *[Report on a scientific expedition to Bohemia and Hungary]; L'Economia rurale nelle Province Lombardo Veneto--rapporto di Luigi Configliachi.* 1835-1836. 71, [10] p., 19 folded leaves of plates: col. ill.; 36 cm.

Two reports on Configliachi's various scientific expeditions. The first indicates that he was accompanied in his travels through Bohemia and Hungary by Beggiato; in Italian and French. Additional material laid in includes 3 A.L.S. On spine: Configliachi - Austria - 1835.

MSS 258 B

737. Copland, James, 1791-1870. *Letter.* 1835. 1 item (1 p.)

A.L.S. (1835 Nov. 4) inviting Jerdan to dinner.

MSS 372 A

738. Crelle, August Leopold, 1780-1855. *Letter.* 1835. 1 item (2 p.)

A.L.S. (1835 June 25); in German.   MSS 380 A

739. Ebillot. *Letter.* 1835. 1 item (1 p.)

A.L.S. (1835 April 26, Hagenau) sent with a work by Hugueny; in French. Formerly laid in Dibner copy of Hugueny's *Nouvelles considerations.*   MSS 468 A

740.   Field, Cyrus West, 1819-1892. *Papers.*
1835-1890. 100 items.

Eighty-four A.L.S. to various correspondents including
Burns, Childs, Estabrook, Hilgard, Landon, Morgan, and
Smith, 9 autograph signatures, 2 A. checks S., 5 D.S.
including several concerned with the New York Society
for the Suppression of Vice.   MSS 509 A

741.   Forbes, Edward, 1815-1854. *Letters.*
[between 1835 and 1854]. 2 items.

A.L.S. (London) to Buckingham declining a speaking
invitation and recommending Lankester, and A.L.S. (June
21) to an unidentified correspondent concerning a collection of specimens including a Lias urchin.   MSS 526 A

742.   Foucault, Léon, 1819-1868. *Papers.*
1835-1862. 5 items: port.

Holograph article (1844 April 1) concerning the movement of leg muscles, and 4 A.L.S., one to Favre; in French.
MSS 530 A

743.   Girard, Philippe Henri de, 1775-1845.
Letters. 1835-1843. 6 items.

Five A.L.S. (one apparently a draft letter) and 1 envelope; in French.   MSS 595 A

**744.   Haldat du Lys, Charles Nicolas Alexan-
dre, 1770-1852.** *Letter.* [1841?]. 1 item (3 p.)

A.L.S. (1841? May 25) to M. Gaiffe, concerning the
Rumkoff apparatus for magnetic polarization; in French.
MSS 652 A

745.   La Rive, Auguste Arthur de, 1801-1873.
Letters. 1835-1856. 3 items.

Three A.L.S. including one to Ampère; in French.
MSS 841 A

746.   *Renseignements sur les opérations topographiques exécutées par le dépôt de la guerre de
Belgique pour la rédaction de la carte du pays.*
[ca. 1850]. 41 leaves, [6] leaves of plates: ill. (some
col.); 36 cm.

Spine title: Opérations Topographiques Belgique.
MSS 284 B

747.   Darwin, Charles, 1809-1882. *Papers.*
1836-1881. 42 items in 4 cases.

Correspondence, including 26 A.L.S. to various correspondents including Busk, Henslow, Hutton, Lubbock,
Mallet, Popper, Stuart-Murray, Westropp, and White,
holograph excerpts from a letter to Chambers, envelopes,
and holograph documents, including pages from the text
of the *Origin of the Species* and *Insectivorous Plants.*
MSS 405 A

748.   Emmons, Ebenezer, 1799-1863. *Letter.*
1836. 1 item (2 p.)

A.L.S. (1836 May 17, Williams College) to Marcy concerning the appointment of a head of the Botany Department and the instruction of farmers about soil use.
MSS 481 A

**749.   Hager, Hans Hermann Julius, 1816-
1897.** *Letter.* [between 1836 and 1897]. 1 item (1 p.)

A.L.S. in German.   MSS 648 A

**750.   Jacquin, Joseph Franz, Freiherr von,
1766-1839.** *Letter.* 1836. 1 item (1 p.)

A.L.S. (1836 April 22); in German.   MSS 747 A

751.   Jahn, Gustav Adolph, 1804-1857. *Letters.* 1836-1845. 2 items.

A.L.S. (1836 July 27, Leipzig) concerning a plan of
mathematical tables, and A.L.S. (1845 April 30, Leipzig)
concerning a meeting of the Astronomical Society at his
observatory; in German.   MSS 748 A

752.   Liebig, Justus, Freiherr von, 1803-1873.
Papers. 1836-1871. 9 items.

Five A.L.S. to Harley and Wolff, 2 D.S., holograph
article "On the nutritive value of extractum carnis," and
a signed title page from an offprint "Die Entwickelung der
Ideen in der Naturwissenschaft"; in German.
MSS 897 A

**753.   Murchison, Roderick Impey, Sir, 1792-
1871.** *Papers.* 1836-1868. 17 items.

D.S., and 16 A.L.S. (two in French) to various correspondents including Herschel, Home, La Rive, and mentioning Milne-Edwards.   MSS 1049 A

754.   Darcet, Jean Pierre Joseph, 1777-1844. *Letter.* 1837. 1 item (1 p.)

A.L.S. (1837 March 4) to an unnamed correspondent referring to requested documents; in French.

MSS 401 A

755.   Elie de Beaumont, Jean Baptiste Armand Louis Léonce, 1798-1874. *Letters.* 1837-1856. 2 items.

A.L.S. (1837 April 18, Paris) to M. Paillon-Boblaye, and A.L.S. (1856 April 17, Paris); in French.

MSS 478 A

756.   Fitzroy, Robert, 1805-1865. *Letters.* 1837-1860. 5 items.

Five A.L.S. to Smyth, Dixon, and unnamed correspondents.

MSS 515 A

757.   Graham, Thomas, 1805-1869. *Letter.* 1837. 1 item (2 p.)

A.L.S. (1837 July 22, Glasgow) to Liebig, concerning Liebig's proposed visit to Glasgow.

MSS 609 A

758. *Magnetism.* 1837. 1 item (40 p.): ill.

MSS 946 A

759.   Pelletier, Pierre Joseph, 1788-1842. *Letter.* 1837. 1 item (2 p.): port.

A.L.S. (1837 July 2, Paris) to Monsieur Ballard introducing Monsieur Schwarzenbert; in German.

MSS 1110 A

760.   Tiedemann, Friedrich, 1781-1861. *Letter.* 1837. 1 item (2 p.)

A.L.S. (1837 Dec. 24?, Heidelberg) to an unnamed friend; in German.

MSS 1468 A

761.   Weber, Wilhelm Eduard, 1804-1891. *Papers.* 1837-1882. 35 items.

Twenty-eight A.L.S. (one with signature cut out) perhaps to Heinrich Weber, 6 holograph leaves, and 1 autograph signature; in German.

MSS 1539 A

762.   Dumas, Jean-Baptiste, 1800-1884. *Letters.* 1838-1879. 6 items.

Six A.L.S. to various correspondents including Peyre and Stenhouse; in French.

MSS 463 A

763.   Hooker, William Jackson, Sir, 1785-1865. *Correspondence.* 1838-1863. 13 items.

Twelve A.L.S. to various correspondents including Cole, Nuttall, and Russell concerning various botanical subjects including Livingstone's African collection and Bedford's interest in Kew Gardens, and an A.L.S. to Hooker.

MSS 718 A

764.   Laurent, Auguste, 1807-1853. *Letter.* [between 1838 and 1853]. 1 item (2 p.)

A.L.S. (undated) to Hofmann concerning Laurent's nucleus theory, containing chemical formulae; in French.

MSS 855 A

765.   Lloyd, Humphrey, 1800-1881. *Letter.* 1838. 1 item (8 p.)

A.L.S. (1838 Feb. 17, Dublin) to Jervis concerning the problems connected with magnetical observations.

MSS 915 A

766.   Mitchell, Maria, 1818-1889. *Letters.* [between 1838 and 1889]. 2 items.

A.L.S. (undated) to her nephew concerning investments and retirement, and A.L.S. (1871 Feb. 19, Poughkeepsie) to Alfred Sterre concerning a lecture which she cannot give.

MSS 1015 A

767.   Ross, James Clark, Sir, 1800-1862. *Letters.* 1838-1852. 3 items.

A.L.S. (1838 Feb. 2, Hyde Park) to an unnamed correspondant concerning Ross's magnetic observations in England, Ireland, and Scotland, and an A.L.S. ([18]52 Aug. 10, [Whitgift?] Hall) with A. envelope, to Ransome congratulating him on receiving the prince's financial support.

MSS 1233 A

768.   Vrain-Lucas, Denis, b. 1818. *Le vorace ou le demon du jeu.* [between 1838 and 1908]. 1 item (14 leaves)

Holograph ms. of an anti-semitic tract; in French.

MSS 1518 A

769. Chevreul, Michel Eugène, 1786-1889.
Papers. 1839-1867. 13 items.

Seven A.L.S. to various correspondents including Castelnau, Guerin, and Sénarmont, and 6 leaves of ms. memoranda; in French.   MSS 339 A

770. Leigh, J. *Letter.* 1839. 1 item (1 leaf)

A.L.S. (1839 Jan. 27, Dresden, Tenn.) to G. W. Palmer & Co. ordering a variety of scientific instruments on letterhead of the *Tennessee Patriot.*   MSS 876 A

771. Brande, William Thomas, 1788-1866.
Letter. 1840. 1 item (1 p.)

A.L.S. (1840 Feb. 3) to Morson containing questions about the properties of amygdalin.   MSS 166 A

772. Chambers, Robert, 1802-1871. *Letters.*
1840-1842. 2 items.

Two A.L.S. (1840 April 13, Edinburgh, and 1842 Sept. 7, St. Andrews) concerning the obtaining of a printing foreman and issues of newspapers.   MSS 326 A

773. Chinon de la Landrière, L. *Letter.* 1840.
1 item (4 p.)

A.L.S. (1840 June 25, Paris) to M. Martineau; in French.   MSS 341 A

774. Coffin, T. G. *Letter.* 1840. 1 item (1 p.)

A.L.S. (1840 Aug. 1, New Bedford) to an unidentified correspondent recommending Audubon.   MSS 357 A

775. Daubeny, Charles Giles Bridle, 1795-1867. *Letters.* 1840. 2 items.

Two A.L.S., one (1840 Mar. 25, Oxford) to David Miller and one (undated) to Mr. Taylor.   MSS 408 A

776. De Morgan, Augustus, 1806-1871. *Letters.* 1840-1869. 9 items.

Nine A.L.S. to various correspondents including Dixon, Rathbone, and Moigno (the last in French).   MSS 427 A

777. Downing, Andrew Jackson, 1815-1852.
Letters. 1840-1850. 2 items.

A.L.S. (1840 Oct. 9, Newburgh) concerning contents of accompanying basket of fruit and flowers, and A.L.S. (1850 Nov. 9, Newburgh) to Mr. Willis concerning the stoves in American schools and Downing's trip to England.   MSS 452 A

778. Ellis, Henry, Sir, 1777-1869. *Letter.*
1840. 1 item (2 p.)

A.L.S. (1840 Dec. 24) to Mr. Mackay concerning the printed catalogue of the Library of the British Museum.   MSS 480 A

779. Gmelin, Leopold, 1788-1858. *Note.*
[1847?]. 1 item (1 p.)

A.N.S.; in German.   MSS 600 A

780. Green, Charles, 1785-1870. *Letter.* 1840.
1 item (1 p.)

A.L.S. (1840 Jan. 24, Highgate) to P. N. Scott with thanks for his copies of newspaper articles on ballon ascensions.   MSS 616 A

781. Griffith, Richard John, Sir, 1784-1878.
Letter. 1840. 1 item (4 p.)

A.L.S. (1840 Sept. 30, Dublin) to Mr. Milne (i.e., Home) concerning Griffith's geological map of Ireland.   MSS 622 A

782. Guilio, Carlo Ignazio, 1803-1859. *Lezioni di meccanica.* [184-?] 304, 323, [23] p.: ill.; 33 cm.

Includes: *Lezioni di dinamica*.   MSS 268 B

783. Kobell, Franz, Ritter von, 1803-1882.
Correspondence. [between 1840 and 1855]. 2 items.

A.L.S. (undated) to Schafhäutl, and A.L.S. (1842 Mar. 16, Munich) from Schafhäutl to John C. Taylor, both regarding English publication of Kobell's *Die Galvanographie;* in German. Letters formerly laid in Dibner copy of *Die Galvanographie.*   MSS 789 A

784.   Lyell, Charles, Sir, 1797-1875. *Letters.* 1840-1871. 11 items: port.

Eleven A.L.S. to various correspondents including Codrington, Geike, Kingsley, and La Hire.   MSS 933 A

**785.   *Memorie scientifiche tratte dalla Biblioteca Italiana e da qualche altro giornale technologico.* [between 1840 and 1845?]. 1 item (46 p.): ill.

Holograph ms. of various journal articles concerning Brunel.   MSS 990 A

786.   Nadar, Félix, 1820-1910. *Letter.* [between 1840 and 1910]. 1 item (1 leaf)

A.L.S. (undated) to M. Wolff concerning (editorial?) proofs; in French.   MSS 1055 A

787.   Petzholdt, Alexander, 1810-1889. *Letter.* 1840. 1 item (2 p.)

A.L.S. (1840 July 29, Dresden), perhaps to Muspratt, on the magneto-electric apparatus the correspondent invented in 1833, but which "Prof. Keil from Munich" is claiming as his own invention, citing Faraday as a witness.   MSS 1128 A

788.   Pfaff, Christian Heinrich, 1773-1852. *Letter.* [between 1840 and 1846]. 1 item (1 leaf)

A.L.S. ([1840?] July 22, Kiel); in German.   MSS 1130 A

789.   Salter, John William, 1820-1869. *Letter.* [between 1840 and 1869]. 1 item (3 p.)

A.L.S. (Aug. 3, Cambridge) to A. Gavan concerning specimens borrowed from the [British] Museum.   MSS 1323 A

790.   Steinheil, Karl August, 1801-1870. *Letters.* 1840-1859. 8 items: ill.

Six A.L.S. (1840-1848, Munich), mostly to Weber, 1 A.L.S. (1855 July 12) possibly to Justus Liebig, and 1 A.L.S. ([1859?] 23, Göttingen) to Wilhelm Steeg. Topics include sight, temperature, and the telescope; in German.   MSS 1413 A

791.   Tonboulie, Monsieur. *Papers.* [between 1840 and 1855]. 2 items: ill.

Two mss. in the same hand: "Le Véloposte" (9 leaves), and a biographical sketch of Tonboulie, author of "Le Véloposte" in French.   MSS 1478 A

792.   Tyndall, John, 1820-1893. *Papers.* [between 1840 and 1893]. 75 items in 3 folders.

Seventy-one A.L.S., two A. envelopes, mostly on letterhead of the Royal Institution of Great Britian, one autograph signature with verse quotation from Emerson, and one ticket blank completed and signed. Various correspondents include Groom Napier, Hilgard, Hooker, Lewes (husband of George Eliot), l'abbé [Moigno], Sir Roderick [Murchison], Playfair, Pollock, and possibly Wallace. In addition to social matters, topics include Eliot, Faraday on magno-electricity, Jones, Tyndall's declining the Edinburgh Chair of Natural Philosophy, thermometers, and experiments with magnetism and electricity. Additional material in third folder.   MSS 1486 A

793.   Volta, Zanino, 1795-1864. *Letter.* 1840. 1 item (1 leaf)

A.L.S. (1840 Mar. 19, Como) to Bolza; in Italian.   MSS 1515 A

794.   Audubon, John Woodhouse, 1812-1862. *Check.* 1841. 1 item (1 p.)

A check signed (1841 Aug. 27, New York).   MSS 30 A

795.   Böttger, Rudolph Christian, 1806-1881. *Letters.* 1841-1842. 2 items.

Two A.L.S. (1841 Dec. 24; 1842 April 15, Frankfurt) to the *Jenaische Allgemaine Literaturzeitung* concerning the reviewing of papers and the publication of a note; in German.   MSS 131 A

796.   Dufrénoy, Ours Pierre Armand Petit, 1792-1857. *Letter.* 1841. 1 item (1 p.)

A.L.S. (1841 Dec. 18) to Cordier concerning a geological map; in French.   MSS 458 A

797. Maceroni, Francis, 1788-1846. *Letter.* 1841. 1 item (3 p.)

A.L.S. (1841 Jan. 14) to the Editor of the *Naval* and *Military Gazette* requesting financial assistance from the readers of the *Gazette.* Printed pamphlet requesting monetary subscriptions in folder. MSS 936 A

798. Talbot, William Henry Fox, 1800-1877. *Letter.* 1841. 1 item (2 p.)

A.L.S. (1841 Sept. 14, Lacock Abbey, Chippenham) to J. W. Gutch concerning reasons for his "failure" [in developing photographs?]: "not using enough of acetic acid, . . .using a sponge instead of a brush; using impure oxidized potassium." MSS 1442 A

799. Waterton, Charles, 1782-1865. *Letters.* 1841-1858. 5 items.

Five A.L.S. to various correspondents including Jupp, Loddiges, and Singleton. MSS 1534 A

800. Crosse, Andrew, 1784-1855. *Letters.* 1842. 2 items.

Two A.L.S. (undated, and 1842 Aug. 25), the former making a dinner appointment with Buckland, the latter recounting Crosse's theory of batteries creating organic life. MSS 384 A

801. Gaultier de Claubry, Henri François, 1792-1878. *Letters.* 1842-1856. 3 items.

A.L.S. (1842 Feb. 11, Paris) to Othon, king of Greece, dedicating a copy of one of his books with draft of the king's reply S. (1843 May 17), and A.L.S. (1856 July 9, Paris) on letterhead of Ecole Superieure de Pharmacie concerning his lectures; in French. MSS 574 A

802. List, Friedrich, 1789-1846. *Letter.* 1842. 1 item (3 p.)

A.L.S. (1842 June 28, Augsberg) to Senator Duckwitz; in German. MSS 906 A

803. Mohl, Hugo von, 1805-1872. *Letter.* 1842. 1 item (1 leaf)

A.L.S. (1842 Dec. 9, Tübingen); in German. MSS 1018 A

804. Robinson, John, Sir, 1778-1843. *Letter.* 1842. 1 item (4 p.): ill.

A.L.S. (1842 Dec. 6, Edinburgh) to Nasmyth concerning the need for precise "observatory instruments," "Mr. Gambey's dividing apparatus," Nasmyth's "thunderbolt hammer," and three of Robison's own inventions: a new method of screw cutting, a "smooth half round file," and "gas blow pipes" for lighting. MSS 1225 A

805. Stroganov, Sergei Grigorevich, graf, 1794-1882. *Papers.* [between 1842 and 1882]. 3 items.

A.L.S. (1878 Aug. 24, Moscow) to "Monsieur le Comte" de Gasparin, and 2 holograph notes S. on title pages of M. Spassky's *Observations météorologiques* (printed in 1842 and 1843); in French. MSS 1426 A

806. Anderson, Thomas, 1819-1874. *Letter.* 1843. 1 item (2 p.)

A.L.S. (1843 Aug. 28, Leith) to Pelouze requesting a recommendation for the professorship of chemistry in Edinburgh. MSS 80 A

807. Bravais, Auguste, 1811-1863. *Letter.* 1843. 1 item (2 p.)

A.L.S. (1843 Oct. 5, Paris) to Erman concerning the need for a French journal which would publish scientific research done in foreign countries, with envelope; in French. MSS 175 A

808. Clanny, William Reid, 1776-1850. *Letter.* 1843. 1 item (2 p.)

A.L.S. (1843 June 21, Wearmouth) concerning his safety lamp. MSS 349 A

809. Dana, James Dwight, 1813-1895. *Letters.* 1843-1861. 8 items.

Eight A.L.S. to various correspondents including Sénarmont and Milne-Edwards. MSS 398 A

810. Dove, Heinrich Wilhelm, 1803-1879. *Papers.* 1843. 4 items.

Two A.N.S., A.D.S. (1843 Aug. 4), and a holograph article S. (12 p.); in German. Additional material includes newspaper obituaries. MSS 451 A

811. Hopkins, William, 1793-1866. *Letters.* 1843-1863. 2 items.

A.L.S. (1843 Oct. 14, Cambridge, England) concerning Professor Roger's papers, and A.L.S. (1863 Sept. 15, Cambridge, England) to Dr. Cranswick concerning Hopkins's paper on geological theories of elevation.   MSS 720 A

812. Mitscherlich, Eilhard, 1794-1863. *Letters.* 1843-1859. 3 items.

Three A.L.S. from Berlin; in German.   MSS 1016 A

813. Poinsot, Louis, 1777-1859. *Letter.* 1843. 1 item (1 leaf)

A.L.S. (1843 July 6, Paris) to an "illustre confrère" thanking him for a gift; in French.   MSS 1155 A

814. Smyth, William, 1797-1868. *Letter.* 1843. 1 item (2 p.)

A.L.S. (1843 Feb. 22, Bowd[oin] Coll[ege]) ordering the construction for the College of "Atwood's Machine for Laws of falling bodies," the "Model of Cycloidal Pendulum," and the "Sliding [Inventice?] Parallelogram, for illustrating the equilibrum of Forces."   MSS 1385 A

815. Strickland, Hugh Edwin, 1811-1853. *Letter.* 1843. 1 item (1 leaf)

A.L.S. (1843 Sept. 29, Cracombe House, Evesham) to an unnamed correspondent concerning the papers of Professor Rogers.   MSS 1425 A

816. Children, John George, 1777-1852. *Letter.* 1844. 1 item (4 p.)

A.L.S. (1844 June 19, Torrington Square) to an unidentified correspondent thanking him for his letters and proofs and hoping they could meet in town.

MSS 340 A

817. Jacobi, Moritz Hermann von, 1801-1874. *Letters.* 1844-1846. 2 items.

A.L.S. (1844 July 3) and A.L.S. (1846 May 30 - June ll, St. Petersburg) to Weber; in German.   MSS 744 A

818. Lassaigne, Jean Louis, 1800-1859. *Letter.* 1844. 1 item (2 p.)

A.L.S. (1844 Dec. 10) to Cottereau concerning antimony; in French.   MSS 843 A

819. Nichol, John Pringle, 1804-1859. *Letters.* 1844-1859. 2 items.

Two A.L.S.: one (1844 Aug. 15) to Home about the need for a "self-registering tide guage(!)" on the River Clyde; the other (n.d.) headed "Observatory--Monday evening," to Mackay, citing Nichol's *Architecture of the Heavens* in response to Mackay's article "Achilae or no Achilae"; also cites Brewster.   MSS 1072 A

820. Page, Charles Grafton, 1812-1868. *Letter.* 1844. 1 item (1 leaf)

A.L.S. (1844 Dec. 14, Patent Office) ordering materials for his "popular course of lectures." Typed transcription included.   MSS 1122 A

821. Raspail, François Vincent, 1794-1878. *Letter.* 1844. 2 items.

A.L.S. (1844 May 29) with A. envelope to Babaud-Laribière, admiring the loyalty shown in his journal and referring to a trail; in French.   MSS 1191 A

822. Rosse, G. H., Sir. *Letters.* 1844-1871. 2 items.

A.L. (1844 [Feb.?] 7, London) declining an invitation to attend "the next anniversary of the 'City of London General Pension Society'" with compliments to Mr. Richardson, and A.L.S. (1871 July 5, Chesham Street) to Mr. Gassiot.   MSS 1235 A

823. Schrötter, Anton, Ritter von Kristelli, 1802-1875. *Papers.* 1844-1878. 11 items.

One holograph ms. S. (1844 July 16): "Ueber die Organization der Lehrfacher für Technische Chemie," 1 partial holograph ms., 2 A.L.S. (1862-1866); 6 A.D.S. (3 dated 1849, 1 each 1866, 1874, 1878), and a handwritten abstract of "Professor Schrötter's Amorphe Phosphor" by August Faberz(?); in German.   MSS 1339 A

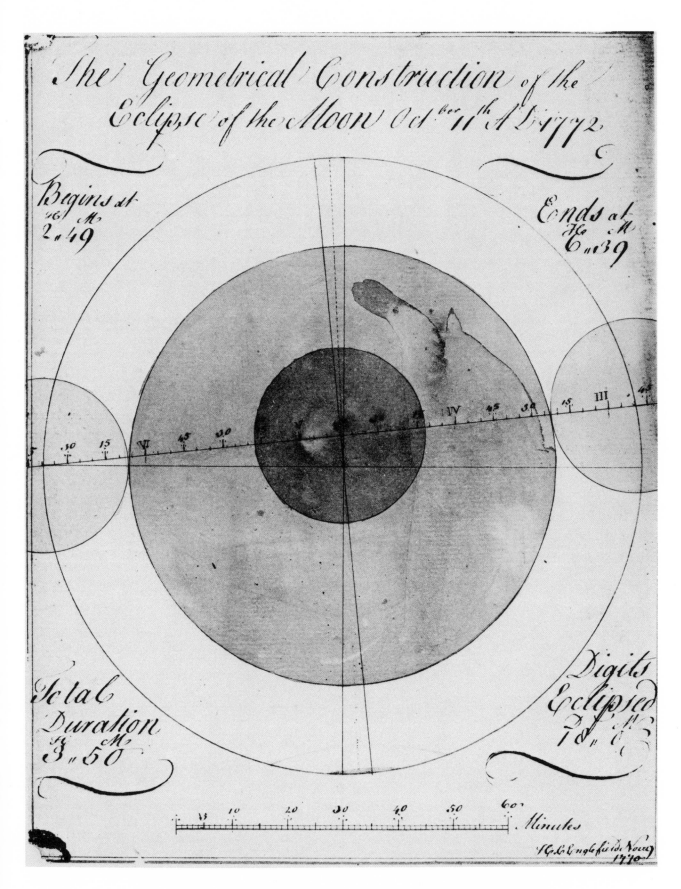

IX.　Drawing of an eclipse of the moon by Sir Henry Englefield
(1770).

Entry No. 282, SI Neg. 84-7187.

824. Champollion-Figeac, Jacques Joseph, 1778-1867. *Letter.* 1845. 1 item (1 p.)

A.L.S. (1845 Jan. 15) concerning a ms. designated "No. 1104"; in French. Additional material includes printed announcements, one with ms. annotations, from the Société des sciences et des arts de Grenoble, some mentioning Champollion jeune. MSS 327 A

825. Combe, George, 1788-1858. *Letters.* 1845-1852. 2 items.

Two A.L.S. (1845 Feb. 20, 1852 March 10, Edinburgh) to publishers (one to Longmans) requesting account sales for Combe's books. MSS 363 A

826. *Descrizione dei telegrafi elettrici art[icol]o tradato ed estratto dall'Eco Francese del l'o Giugno 1845.* 1845. 2 items.

Rough copy (7 leaves) and fair copy (5 leaves). MSS 432 A

827. Erlenmeyer, Richard August Carl Emil, 1825-1909. *Invitation.* [between 1845 and 1909]. 1 item (1 p.)

A. invitation S.; in German. MSS 488 A

828. Ferrucci, Antonio. *Letters.* 1845-1851. 6 items.

Six A.L.S. to Bertagnini; in Italian. MSS 508 A

829. Jackson, Charles Thomas, 1805-1880. *Directions for the construction of an improved mountain barometer.* 1845. 1 item (10 p.)

A. ms. S. (1845 Mar. 14, Boston). MSS 741 A

830. Labarraque, Antoine Germain, 1777-1850. *Letter.* 1845. 1 item (3 p.)

A.L.S. (1845 Nov. 22, Paris) to the Minister of Public Instruction concerning the meeting of a medical congress; in French. MSS 807 A

831. Palmieri, Giuseppe, 1721?-1794? *Letter.* 1845. 1 item (1 p.)

A.L.S. (1845) to Signor Cavaliere Lobetti; in Italian. MSS 639 A

832. Poggendorff, Johann Christian, 1796-1877. *Letter.* 1845. 1 item (2 p.)

A.L.S. ([18]45 Jan. 20, Berlin) to Barth; in German. MSS 1153 A

833. Rosse, William Parsons, 3rd earl of, 1800-1867. *Letters.* 1845-1864. 3 items.

Three A.L.S., one (1864 June 16) supposedly to Yates concerning a meeting of the Royal Society and support for a metrical system. MSS 1097 A

834. Sowerby, James de Carle, 1787-1871. *Letter.* 1845. 1 item (2 p.)

A.L.S. (1845 Feb. 19, Botanic Gardens, [London]) to Edward Charlesworth concerning Sowerby's and Dr. Mantell's works on fossils. MSS 1395 A

835. Berghaus, Heinrich Karl Wilhelm, 1797-1884. *Tafeln zur statistik der österreichischen monarchie für das Jahr 1842.* 1846 1 item (2 p.)

Holograph ms. S. (1846 Nov. 20) in German. MSS 91 A

836. Clapp, Henry, 1814-1875. *Letter.* 1846. 1 item (2 p.)

A.L.S. (1846 Sept. 2, London) concerning the acquiring of "a million or two of dollars," and promising to send letters of more interest than those in the *Pioneer*. MSS 351 A

837. Dase, Zacharias, 1824-1861. *Papers.* 1846. 3 items.

A.L.S. (1846, Berlin), 1 p. of mathematical calculations in pencil, and an autograph signature; in German. MSS 406 A

838. Geinitz, Hanns Bruno, 1814-1900. *Letter.* 1846. 1 item (4 p.)

A.L.S. (1846 Feb. 3) concerning a festival meeting of the Gerverbeverein; in German. MSS 583 A

839. **Gould, Benjamin Apthorp, 1824-1896.** *Letters.* 1846-1848. 4 items.

Four A.L.S., 3 to Militzer; in German.   MSS 605 A

840. **Kendall, Amos, 1789-1869.** *Letter.* 1846. 1 item (2 p.)

A.L.S. (1846 May 10, Washington, D.C.) to Marcy concerning the building of a telegraph line from Washington to New Orleans by the government or by Morse.
   MSS 780 A

841. **Le Verrier, Urbain Jean Joseph, 1811-1877.** *Letters.* 1846-1876. 27 items.

Twenty-six A.L.S. and 1 L.S. in a fair hand to various correspondents including Barral, Bernays, and Lubbock; in French.   MSS 887 A

842. **Markoe, Francis.** *Letter.* 1846. 1 item (4 p.)

A.L.S. (1846 June 24, Washington) to H. P. Sartwell, concerning the *Bulletin of the National Institute for the Promotion of Science.*   MSS 961 A

843. **Valenciennes, Achille, 1794-1865.** *Letter.* 1846. 1 item (1 leaf)

A.L.S. (1846 Oct. 24) concerning live and fossilized specimens of sea urchins; in French.   MSS 1494 A

844. **Herschel, August Wilhelm Eduard Theodor, 1790-1856.** *Letter.* 1847. 1 item (4 p.)

A.L.S. (1847 Sept. 25 Breslau); in German. Additional sheets of biographical notes.   MSS 689 A

845. **Neeff, Christian Ernst, 1782-1849.** *Letter.* 1847. 1 item (1 leaf)

A.L.S. (1847 May 6, Frankfurt); in German.
   MSS 1065 A

846. **Petzval, Joseph, 1807-1891.** *Integration der Differential-Gleichungen von Linearen Form.* 1847. 1 item (84 p.)

A. ms. S. (1847 Mar. 6, Vienna), part one of a text on differential equations to be published by Haidinger. Last page has signed note (1847 June 5) by Haidinger and marginal note (1847 June 12), roman numeral XVI precedes title; in German.   MSS 1129 A

847. **Adams, John Couch, 1819-1892.** *Letters.* 1848-1855. 2 items.

Two A.L.S., one (1848 Oct. 21, Cambridge) to Rev. Albert Alston concerning the election of one Bateson to the public oratorship of Cambridge, and one (1855 June 15, Greenwich) to Rev. John Davies concerning the latter's article on Semitic languages.   MSS 64 A

848. **Antrobus, Edmund, 1818-1899.** *Letters.* 1848. 2 items.

A.L.S. (1848 Sept. 23, Amesbury) from Antrobus to unidentified concerning a candidate for assistant surgeon at Middlesex Hospital, W. I. Anderson. A.L.S. (18[--?] July 10, London) from Antrobus to Mr. Thomas concerning Antrobus's inability to do anything but compilation anymore.   MSS 10 A

849. **Bunsen, Robert Wilhelm Eberhard, 1811-1899.** *Letters.* 1848-1888. 14 items; port.

Thirteen A.L.S. and one envelope to various correspondents including Meyer, Struve, and Schielt; in German.   MSS 203 A

850. **Ferguson, James, 1797-1867.** *Letter.* 1848. 1 item (3 p.)

A.L.S. (1848 Dec. 9, Washington) to Rev'd George Bush requesting that he review Ferguson's article on the Coast Survey and help find a publisher for it.
   MSS 1599 A

851. **Flourens, Marie Jean Pierre, 1794-1867.** *Letters.* 1848-1862. 4 items.

Four A.L.S., two on letterhead of the Institut de France; in French.   MSS 521 A

852. **Graves, Robert James, 1796-1853.** *Letter.* 1848. 1 item (3 p.)

A.L.S. (1848 June 23) to William Wilde concerning Graves' trip in Connaught.   MSS 613 A

853. **Hind, John Russell, 1823-1895.** *Letters.* 1848-1850. 3 items.

Three A.L.S. to Lubbock concerning astronomical subjects.   MSS 704 A

854. Latham, Robert Gordon, 1812-1888.
Letter. [between 1848 and 1888]. 1 item (2 p.)

A.L.S. (undated, New Malden, Kingston-on-Thames, Surrey) to an unnamed correspondent concerning one of Latham's writings on the languages of Africa.

MSS 847 A

855. Agassiz, Louis, 1805-1873. *Drawings illustrating the Lowell Lectures on comparative embryology.* [ca. 1849] [10] leaves: all col. ill.

Laid in: contemporary envelope with title, probably in Mrs. Agassiz's hand; correspondence between donor and dealer; photostats of two letters from Agassiz.

MSS 1 B

856. Barrande, Joachim, 1799-1883. *Letter.* 1849. 1 item (1 p.)

A.L.S. (1849 Aug. 25) to Sandberger; in German.

MSS 49 A

857. Billroth, Theodor, 1829-1894. *Note.* [between 1849 and 1894]. 1 item (1 p.)

Calling card with holograph note to Dr. R. Török; in German.

MSS 114 A

858. Bright, John, 1811-1889. *Letter.* 1849. 1 item (4 p.)

A.L.S. (1849 April 24, London) to Osborne-Morgan concerning Locke's bill permitting passenger railroad trains on Sundays.

MSS 178 A

859. Ericsson, John, 1803-1889. *Papers.* 1849-1885. 15 items: ill.

Twelve A.L.S., 1 A.L., and 1 A.D.S., with drawing, applying for patent for a caloric engine. Additional material includes 4 holograph letters referring to Ericsson.

MSS 487 A

860. Hofmann, August Wilhelm von, 1818-1892. *Papers.* 1849-1891. 13 items.

Ten A.L.S., one in English to Playfair and one condoling Wurtz's sister on his death, one A. note S., one A. envelope, and one calling card addressed in French to Paalzow; in German.

MSS 710 A

861. Hooker, Joseph Dalton, 1817-1911. *Letters.* 1849-1888. 11 items.

Eleven A.L.S. to various correspondents including Cole and Cooper.

MSS 716 A

862. Mayer, Julius Robert von, 1814-1878. *Papers.* 1849-1870. 4 items: port.

Two A.L.S., 1 A.S. with epigram, and 1 A. ms. "Die Toricellische Leere von J. R. Mayer in Heilbronn" (2 p.); in German.

MSS 983 A

863. Seidel, Philipp Ludwig von, 1821-1896. *Letters.* 1849-1873. 4 items.

Four A.L.S.: two (1849, 1857) to Militzer, one (1873) to a colleague, and one to Professor D. A. Vogel; in German.

MSS 1353 A

864. Smyth, Charles Piazzi, 1819-1900. *Letters.* 1849-1889. 4 items.

A.L.S. (1849 Apr. 21) to S. John concerning articles on shooting stars, A.L.S. (1861 Nov. 15) to Lowell Reese on letterhead of the Royal Observatory, Edinburgh, and 2 A.L.S. (1889 June 20 and July 22, Clova, Ripon) to Knight concerning astronomy.

MSS 1383 A

865. Cauchy, Augustin Louis, baron, 1789-1857. *Papers.* 1822-1831. 4 items.

Three A.L.S. and a holograph ms. titled: "Résumé d'un mémoire sur la mecanique celeste et sur un nouveau calcul, appellé calcul des limites"; in French. Additional material includes ms. biography of Cauchy.

MSS 314 A

866. Bartoli, Adolfo, 1833-1894. *Card.* [between 1853 and 1894]. 1 item.

Visiting card with autograph message on recto and mathematical calculations on verso; in Italian.

MSS 27 A

867. Beaufort, Francis, Sir, 1774-1857. *Letter.* 1850. 1 item (2 p.)

A.L.S. (1850 May 14) from Beaufort to Lubbock concerning Lubbock's gnomonic projection.

MSS 25 A

868. Brunel, F. *Letter.* 1850. 1 item (1 p.)

A.L.S. (1850 May 29, Dunkirk) to M. Renouard stating which train Brunel will take for his visit; in French.

MSS 559 A

869. Burmeister, Hermann, 1807-1892. *Letter.* 1850. 1 item (2 p.)

A.L.S. (1850 May 19) to Wigand; in German.

MSS 205 A

870. Chasles, Michel, 1793-1880. *Letters.* 1850-1880. 12 items.

Twelve A.L.S. to various correspondents including Lamé and possibly Sénarmont; in French.   MSS 336 A

871. Christie, Samuel Hunter, 1784-1865. *Correspondence.* 1850. 4 items.

A.L.S. (1850 Nov. 4) to Brande asking which paper on an enclosed list is most worthy of the Royal Medal in Chemistry; and 2 draft responses on the same sheet, presumably by Brande, asking whether Christie's request is official or personal, and recommending no award be given.

MSS 344 A

872. Claudianus, Claudius, ca. 370-ca. 404. *L'aimant extrait des Idylles de Claudien traduit par M. Souquet de la Tour.* [after 1850]. 1 item (3 p.)

Autograph ms. containing a French translation of a poem about the magnet included in a 4th century work by Claudianus.   MSS 926 A

873. Gauss, Karl Friedrich, 1777-1855. *Des formes quadratiques [par] Gauss; Premiers éléments de la théorie des nombres par U[mberto] Scarpis professeur au lycée Scripioné [Maffei?] à Verone.* [ca. 1860]. 24, [61] leaves; 36 cm.

Bound ms. with two mathematical texts beginning at opposite ends of codex, with table of contents (4 leaves) for Gauss' work in pocket inside back cover; 11 blank leaves between texts; in French. Spine title: Nombres.

MSS 1277 B

874. Maury, Matthew Fontaine, 1806-1873. *Letters.* 1850-1855. 3 items.

Three A.L.S., one (1850 Sept. 29) to Harper & Bros. concerning deep sea soundings and submarine telegraph, and two (1854 Dec. 7 and 1855 Nov. 21) to Meriam concerning benefits for Kane.   MSS 977 A

875. Meyer, Thomas Lothar. *Letter.* [between 1850 and 1900]. 1 item (3 p.)

A.L.S. (undated) to Lascelles; in German.

MSS 999 A

876. Pelouze, Théophile Jules, 1807-1867. *Letter.* 1850. 1 item (1 leaf)

A.L.S. ([18]50 June 26, Paris) to Monsieur Michel about Monsieur Nicot's standing in a competition; on letterhead of the Commission des Monnaies et Médailles; in French.   MSS 1111 A

877. Sheepshanks, Richard 1794-1855. *Letters.* 1850-1851. 3 items.

Two A.L.S. (1850 July 3 and 1851 June 23, Royal Astronomical Society) to Lubbock, and 1 A.L.S. to Airy of the Royal Observatory at Greenwich.   MSS 1364 A

878. Argelander, Friedrich Wilhelm August, 1799-1875. *Letter.* 1851. 1 item (3 p.)

A.L.S. (1851 May 6) from Argelander to unidentified.

MSS 15 A

879. Bauernfeind, Karl Maximilian von, 1818-1894. *Letters.* 1851-1856. 2 items.

Two A.L.S. (1851 Nov. 22 and 1856 Sept. 24); in German.   MSS 52 A

880. Cole, Henry, Sir, 1808-1882. *Letter.* 1851. 1 item (1 p.)

A.L.S. (1851 Feb. 10) to Lewis concerning the covering of a monolith.   MSS 361 A

881. Du Chaillu, Paul Belloni, 1831-1903. *Letter.* [between 1851 and 1903]. 1 item (3 p.)

A.L.S. (Edinburgh) thanking Owen for defending him in *The Athenaeum* against Gray's attacks.   MSS 323 A

882. Harrison, Frederic, 1831-1923. *Letter.* [between 1851 and 1923]. 1 item (4 p.)

A.L.S. (Nov. 8) to Cunningham concerning a paper by Wallace to be printed.　　MSS 666 A

883. Hauer, Franz, Ritter von, 1822-1899. *Letter.* 1851. 1 item (1 p.)

A.L.S. (1851 Mar. 12); in German.　　MSS 672 A

884. Heisenberg, E. *Letters.* 1851-1857. 3 items.

Three A.L.S.; in German.　　MSS 679 A

885. Herschel, John Frederick William, Sir, 1792-1871. *Papers.* 1851-1869. 53 items.

Thirty-five A.L.S. to various recipients including Arago, Collingwood, Daniell, Elliott, Lubbock, Moigno, Quetelet, and Schumacher, 2 autograph signatures, 1 D.S., 8 holograph mss., 4 envelopes, 2 printed sheets with annotations, and a ms. copy of a speech by Herschel.　　MSS 695 A

886. Lassell, William, 1799-1880. *Letters.* 1851-[1860?]. 2 items.

A.L.S. (1851 Nov. 13, Starfield) declining invitation from Lubbock, and A.L.S. (Wednesday, Bradstones) declining invitation from Mrs. Thompson. Newspaper obituary included.　　MSS 846 A

887. Nott, Eliphalet, 1773-1866. *Letter.* 1851. 1 item (1 leaf)

A.L.S. (1851 June 19, Union College) to James R. Westcott concerning Dr. Benedict of Utica.　　MSS 1079 A

888. Schleiden, Matthias Jacob, 1804-1881. *Papers.* 1851-1881. 3 items.

A.L.S. ([18]53 Apr. 6, Jena) to an unknown correspondent, A.L.S. (1851 Aug. 15, Jena) to Clemens mentining Cohn, and A. quotation from Goethe, S. (1881 Feb. 6, Wiesbaden) on printed page of the "Autographen-Album des Deutschen Reichs"; in German.　　MSS 1332 A

889. Wallich, Nathaniel, 1786-1854. *Letter.* 1851. 1 item (3 p.)

A.L.S. (1851 July 9, London) to R. Westinhalz on letterhead of the Great Exhibition of the Works of Industry of all Nations, 1851.　　MSS 1528 A

890. Anschütz, Richard, 1852-1937. *Autograph list.* 1852. 1 item (1 p.); port.

Holograph itemized list signed; in German.
　　MSS 81 A

891. Archer, Frederick Scott, 1813-1857. *Letter.* 1852. 1 item (3 p.)

A.L.S. (1852 March 26, London) from Archer to Dr. Diamond requesting samples of the collodion process from Diamond to send to Paris.　　MSS 12 A

892. Du Bois-Reymond, Emil Heinrich, 1818-1896. *Letters.* 1852-1893. 8 items: ill.

Eight A.L.S., one to Virchow and others referring to Meyer and Berthold; in German.　　MSS 455 A

893. Fitch, Joshua Girling, Sir, 1824-1903. *Letter.* 1852. 1 item (3 p.)

A.L.S. (1852 Oct. 2, London) to Tyndall with a copy of Fitch's book and warmest regards and thanks.
　　MSS 514 A

894. Kirchhoff, Gustav Robert, 1824-1887. *Papers.* 1852-1880. 10 items.

Seven A.L.S., and 1 annotated page from a desk calendar, in German, 2 autograph signatures with Latin epigrams.　　MSS 784 A

895. Loomis, Elias, 1811-1889. *Letter.* 1852. 1 item (1 leaf)

A.L.S. (1852 Oct. 11, New York University) concerning his contributing to a new journal.　　MSS 921 A

896. Owen, Richard, Sir, 1804-1892. *Letters.* 1852-1888. 14 items.

Fourteen A.L.S., including one in French (1855 Dec. 27, London) requesting that "Monseigneur" thank Napoleon III for awarding Owen the "Decoration de Chevalier de la Legion d'Honneur" and one (1852 May 26, [British] Museum) to Balfour. Other recipients include Lyell, Sir Joseph, and McCormick; in English.   MSS 1152 A

897. Playfair, Lyon Playfair, 1st baron, 1818-1898. *Letters.* 1852-1891. 8 items.

Eight A.L.S. to various correspondents including N. E. Mitchell, James L. Ambrose, J. E. Ashworth, and Colonel Biddulph. The letter to Biddulph concerns "recyclable gas."   MSS 1148 A

898. Plücker, Julius, 1801-1868. *Letters.* 1852-1859. 2 items.

A.L.S. (1852 Oct. 10, Paris) to Moigno, and A.L.S. (1859 Dec. 21, Bonn) on optical and magnetic tests on mica; in French.   MSS 1150 A

899. Balard, Antoine Jerome, 1802-1876. *Letters.* 1853-1860. 8 items.

Seven A.L.S. (only 1 dated 1860 Oct. 26) from Balard to Despretz, Chevreul, Raynier and unidentified addressees; 1 A.D.S. (1853 May 10, Paris): certificate of attendance for Dr. Zabani; in French. Finding aid in folder.   MSS 41 A

900. Bert, Paul, 1833-1866. *Letter.* [between 1853 and 1866]. 1 item (1 p.)

A.L.S. (undated) on letterhead of the Chambre des députés; in French.   MSS 99 A

901. Bonney, Thomas George, 1833-1923. *Minerals.* [between 1853 and 1923]. [232] p.; 18 cm.   MSS 225 B

902. Brodie, Benjamin Collins, Sir, 1783-1862. *Letter.* 1853. 1 item (2 p.)

A.L.S. (1853 Sept. 16, Surrey) concerning his inability to see Lord Abingdon.   MSS 183 A

903. Büchner, Ludwig, 1824-1899. *Letters.* 1853. 2 items.

Two A.L.S. (1853 July 6, Tübingen; and 1853 Oct. 15, Darmstadt); in German.   MSS 198 A

904. Burckhardt, Johann Karl, 1773-1825. *Papers.* 1853. 3 items.

A.L.S. and 2 pages of calculations entitled "Mondsgleichungen durch mittlere argumenti nach meiren tafeln," and "Formule de la Longitude lunaire dans l'ordre des mes nouvelles tables"; in German and French.   MSS 204 A

905. Chantepie. *Drawing and explanation.* 1853. 1 item (1 leaf folded in four)

Drawing of axle and wheels (1853 Feb. 19, Chatellerault) with ms. explanation signed; in French.   MSS 329 A

906. Fairbairn, William, Sir, 1789-1874. *Letters.* 1853-1872. 5 items.

Four A.L.S. and a holograph sheet of notes (2 p.). Additional material inculdes a newspaper obituary.   MSS 498 A

907. Gallaudet, Thomas, 1822-1902. *Letter.* 1853. 1 item (1 p.)

A.L.S. (1853 Nov. 9, New York) to Stephen Warren, requesting a contribution for St. Anns Church for Deaf-Mutes.   MSS 564 A

908. Gorup von Besanez, Eugen Franz Seraphin, Freiherr von, 1817-1878. *Letter.* 1853. 1 item (1 p.)

A.L.S. (1853 Aug. 7, Erlangen) to J. Pohl, possibly Joseph Johann Pohl; in German.   MSS 604 A

909. Mädler, Johann Heinrich von, 1794-1874. *Papers.* 1853-1865. 8 items.

Five A.L.S. to various correspondents including Braniss, Müller, and Zeune, 1 A. ms. S., 1 A. ms., and 1 envelope; in German.   MSS 944 A

910. Ohm, Georg Simon, 1787-1854. *Letter.* 1853. 1 item (1 leaf)

A.L.S. (1853 Nov. 19) to Lamont; in German.   MSS 1084 A

911. Pontécoulant, Philippe Gustave Le Doulcet, comte de, 1795-1874. *Letter.* 1853. 1 item (2 p.)

A.L.S. (1853 Nov. 10, Paris) to Lubbock about Lubbock's research on the moon and the astronomical topics Pontécoulant plans to treat in his next book, with envelope; in French. MSS 1160 A

912. Reichenbach, Heinrich Gottlieb Ludwig, 1793-1879. *Letters.* 1853-1873. 2 items.

A.L.S. (1853 Aug. 21, Dresden) to Nicard mentioning Verreaux, and an A.L.S. (1873 May 20) on letterhead of the Academia Caesarea Germanica Leopoldino-Carolina citing Petersen's *Wiesenbaurythem;* in German. MSS 1199 A

913. Farmer, W.(?) *Letter.* [1854?]. 1 item (2 leaves)

A.L.S. ([1854?] May [1?]) to Lacelles (possibly F. H. Lascelles) on letterhead of the Royal College of Science, London, concerning a specimen which "Specialists pronounce. . .to be the [fossil] canine of a horse" and which Farmer has shown to Howes and sent to "Newton who is the chief authority on these things." MSS 1598 A

914. Nasmyth, James Hall, 1808-1890. *Letters.* 1854-1887. 8 items.

Seven A.L.S. and 1 envelope. Additional material includes obituaries and biographical clippings. MSS 1058 A

915. *Notes on Ibbertson's Geometric Chuck by Holtzapffel & Co.* 1854. 30 p.: ill.; 23 cm.

Explaination and illustrations concerning the principles and use of the chuck, including a table of settings. Bound with Ibbetson's "A brief account of Ibbetson's geometric chuck. . ." published in London (1833). MSS 1609 B

916. Pasteur, Louis, 1822-1895. *Papers.* 1854-1889. 8 items.

Two A.L.S., 3 A. cards S., 1 business card, 1 A. ms. (1 p.) on a note by Geoffroy Saint-Hilaire, and 1 A. ms. S. (8 p.) "Sur le dimorphisme dans les substances actives"; in French. [XIII, XIV B] MSS 1100 A

917. Ramsay, Andrew Crombie, Sir, 1814-1891. *Letters.* 1854-1877. 3 items.

Three A.L.S. (1854 May 13, [1870 July?], and 1877 Dec. 4), two on letterhead of the Geological Survey of the United Kingdom, to Trimmer (about glacial theory), Crosse, and Baily. MSS 1186 A

918. Sprengel, Hermann, 1834-1906. *Letters.* [between 1854 and 1906]. 2 items: ill.

A.L.S. (1900 Nov. 3, London) to Thompson concerning a lengthy letter to the President and Council of the Chemical Society, with Thompson's holograph personal memo signed with initials, on the verso, and conclusion (2 p.) of A.L.S. to Dr. Stenhouse concerning a pump. MSS 1401 A

919. Stokes, George Gabriel, Sir, 1819-1903. *Letters.* 1854-1879. 4 items.

Four A.L.S. to various correspondents including H. Alfred [Cummington?], Pole, and Sénarmont. MSS 1422 A

920. Universal Electric Telegraph Co. *Deed of settlement of the Universal Electric Telegraph Company.* 1854. [78] p.; 45 cm.

Document (1854) to establish the Universal Electric Telegraph Company, naming John Walker Wilkins as principal subscriber, Andrew Caldecott as party of the second part, and Frederick Burmester as party of the third part. Six directors and 4 auditors are named. No signatures; 39 blank pages, some with seals, for subscribers' signatures. Spine title: Deed--Universal Elec. Telegraph Co. MSS 1302 B

921. Becquerel, Edmond, 1820-1891. *Letters.* 1855-1890. 11 items.

Ten A.L.S. including one to Pingard (with envelope) and one to Maindron mentioning Becquerel's son Henri; in French. MSS 59 A

922. Cornell, Ezra, 1807-1874. *Papers.* 1855-1871. 2 items.

A.L.S. (1855 May 28, Ithaca) recommending Mr. Hopkins as a telegraph operator, and a signed check (1871 June 17) to J. P. Allen for mastodon remains. MSS 374 A

923. Franklin, Jane Griffin, Lady, 1792-1875. *Papers.* 1855-1857. 5 items.

Four A.L.S., one to Ross, and a handwritten calling card. MSS 541 A

924. Grove, William Robert, Sir, 1811-1896. *Letters.* 1855-1888. 2 items.

A.L.S. (1855 Sept. 24, Etretat) to M. Du Moncel with thanks for hospitality and information about experiments on iron bars, and A.L.S. (1888 May 24) concerning an investment. MSS 626 A

925. Kane, Elisha Kent, 1820-1857. *Letter.* 1855. 1 item (1 p.)

A.L.S. (1855 Nov. 16, Philadelphia) to E. Meriam concerning the keeping of a meterological record. MSS 766 A

926. Kelvin, William Thomson, baron, 1824-1907. *Letters.* 1855-1906. 41 items: ill., some col.

Twenty-six A.L.S. (1855-1904), 5 A. envelopes (1882-1902), 5 L.S. (1883-1906), 2 T.L.S. (1905), 1 A. postcard S. (1905), 1 autograph signature, and 1 colored illustration with formulas. Correspondents include Blackburn, Carhart, R. C. Dudley, Flower, and the Managing Committee of The International Electrical Exhibition of Vienna (1883). Topics include Berthelot, Marconi, Watt, a multi-celled battery, a galvanometer, and the measurement of electricity and magnetic force. MSS 1464 A

927. Lovering, Joseph, 1813-1892. *Notes from Prof. Lovering's lectures on electricity.* 1855-1856. 81, [18] p.: ill.; 13 cm.

Apparently incomplete notes from lectures given Mar. 15-May 25, 1855. Includes "Hints from Prof. Lovering" (18 p.), 1856. Label of Eayrs & Fairbanks, Account Book Manufacturers, with "J. F. Flagg" printed on, laid inside front cover. MSS 576 B

928. Moigno, François Napoléon Marie, 1804-1884. *Letters.* 1855-1869. 3 items.

Three A.L.S.; in French. MSS 1019 A

929. Sharpe, Daniel, 1806-1856. *Letter.* 1855. 1 item (3 p.)

A.L.S. (1855 Nov. 16, Soho Square, [London]) to Prestwich identifying and discussing "the fragments of organic remains from the Artesian Well on Highgate Hill." MSS 1362 A

930. Wislicenus, Johannes, 1835-1902. *Letter.* [between 1855 and 1902]. 1 item (1 leaf)

A.L.S. in German. MSS 1578 A

931. Bertrand, Joseph Louis François, 1822-1900. *Letters.* 1856-1898. 3 items.

Three A.L.S., one (1856 Feb. 18) applying for a seat in the Académie, one (1898 Oct. 7) to Pingard concerning a payment to Mme. Serret, and one ([1886?] Aug. 24) concerning the centenary of the Académie; in French. MSS 102 A

932. Huxley, Thomas Henry, 1825-1895. *Papers.* 1856-1894. 68 items: port.

Sixty-four A.L.S. to various correspondents including Barlow, Becker, Cole, Evans, Farrar, Garnett, Günther, Hart, Milne-Edwards, Spring Rice, Tyndall, and Wallace, at least one mentioning Darwin; A.S. with epigram on place of women; another A.S.; envelope; and portrait. MSS 738 A

933. Lockyer, Joseph Norman, Sir, 1836-1920. *Papers.* [between 1856 and 1920]. 5 items.

Three A.L.S., one A. postcard signed with initials, and one A. envelope S. MSS 916 A

934. Thénard, Arnaud Paul Edmond, baron, 1819-1884. *Letters.* 1856-1857. 3 items.

Three A.L.S. to Sénarmont concerning the Société des amis des Sciences, one requesting him to write an official letter to "M. Paul," another recommending M. Pingard; in French. MSS 1453 A

935. Wood, Charles. *Letter.* 1856. 1 item (3 p.)

A.L.S. (1856 Mar. 4) to an old acquaintance explaining that he was unable to grant his friend's wishes. MSS 1585 A

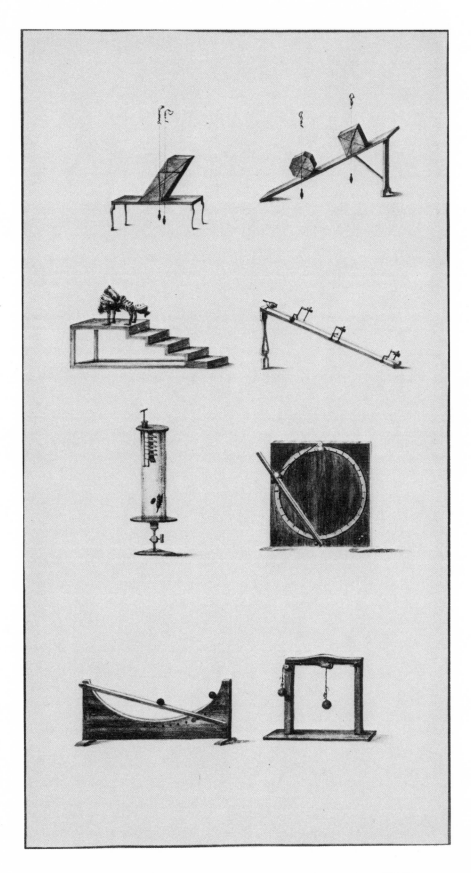

X. Equipment for use in physical experiments (ca. 1800).
Entry No. 286, SI Neg. 84-7191.

936. Frankland, Edward, Sir, 1825-1899. *Letters.* 1857-1882. 5 items.

Five A.L.S. to various correspondents including Mills.

MSS 537 A

937. Haidinger, Wilhelm Karl, Ritter von, 1795-1871. *Letter.* 1857. 1 item (3 p.)

A.L.S. (1857 May 17, Vienna); in German.

MSS 651 A

938. Hunt, Thomas Sterry, 1826-1892. *Letters.* 1857-1859. 4 items.

Four A.L.S. to Sénarmont; in French.   MSS 736 A

939. Muspratt, James Sheridan, 1821-1871. *Letter.* 1857. 1 item. (2 p.)

A.L.S. (1857 April 27, Liverpool) to an editor concerning an important discovery by S. [Medlock?] on letterhead of the Royal College of Chemistry.   MSS 1054 A

940. Philosophical Club of London. *Attendance list and minutes.* 1857. 1 item (2 p.)

List of signatures of members attending the meeting (1857 Feb. 5) with a rough draft of the minutes of that meeting on the verso.   MSS 919 A

941. Piria, Raffaele, 1815-1865. *Correspondence.* 1857. 1 item (4 p.)

A.L.S. (1857 Sept. 22, Paris) to Matteucci about finding a job for Tassinari; with Matteucci's letter to Tassinari on last two pages; in Italian.   MSS 1143 A

942. Riemann, Bernhard, 1826-1866. *Letters.* 1857-1866. 6 items.

Six A.L.S., mostly written at Göttingen, one of them (1857 July 7, Bremen) mentioning Dedekind; in German.

MSS 1214 A

943. Secchi, Pietro Angelo, 1818-1878. *Letters.* 1857-1872. 5 items.

One A.L.S. (1864 Apr. 30) to Moigno concerning electricity and 4 A.L.S. (1857-1872): one to Ferrari concerning the visit of the "famiglia odescalchi," one to Monsignore [Nardi?], one to Respighi concerning the telegraph, and one to P. Egidi; in Italian.   MSS 1347 A

944. Waals, Johannes Diderik van der, 1837-1923. *Papers.* [between 1857 and 1935]. 3 items: ill.

Holograph notebook in Dutch containing equations, tables, diagrams, and chemical notation, and A.L.S. with A. envelope S. (1935 Mar. 28, Amsterdam) in German from Waals, Jr., to Armin Weiner concerning the notebook.   MSS 1603 A

945. Béchamp, Pierre Jacques Antoine, 1816-1908. *Letters.* 1858-1903. 2 items.

Two A.L.S. (in French), one (1858 Nov. 22, Montpellier) requesting the unidentified recipient to recommend Saintpierre to the Académie.   MSS 55 A

946. Gunton, William. *Letter.* 1858. 1 item (4 p.)

A.L.S. ([18]58 Feb. 3, Washington) to J. M. Carlisle critiquing his court case in which he alleged fraud on the part of an inventor. Gunton details the statistics of wood burning aboard the *Columbia* before and after the fraudulant invention was used.   MSS 1600 A

947. Magnus, Heinrich Gustav, 1802-1870. *Letter.* 1858. 1 item (4 p.)

A.L.S. (1858 Feb. 12, Berlin) to Wöhler; in German.

MSS 949 A

948. Maxwell, James Clerk, 1831-1879. *Letters.* 1858-1878. 3 items.

Three A.L.S., one to Tyndall.   MSS 981 A

949. Otto, Friedrich Julius, 1809-1870. *Letter.* 1858. 1 item (2 p.)

A.L.S. ([18]58 Feb. 22); in German.   MSS 1151 A

950. Rühmkorff, Heinrich Daniel, 1803-1877. *Letters.* 1858-1860. 2 items.

A.L.S. (1858 Nov. 27, Paris) in French concerning a bill, and A.L.S. (1860 Sept. 4, Paris) in German with postscript mentioning Dr. Alexandre.   MSS 1243 A

951. Welsh, John 1824-1859. *Letter.* 1858. 1 item (2 p.)

A.L.S. (1858 Apr. 30, Kew Observatory, Richmond) inviting the unnamed correspondent to visit the observatory.   MSS 1553 A

952. Boncompagni-Ludovisi, Baldassare, principe, 1821-1894. *Letters.* 1859-1867. 3 items.

Two A.L.S. (1859 April 22; and 1860 Mar. 24); in Italian, and L.S. (1867 June 21); in French.

MSS 139 A

953. Clausius, Rudolf Julius Emmanuel, 1822-1888. *Letter.* 1859. 1 item (3 p.); ports.

A.L.S. (1859 April 8, Zurich) to Barth with corrections for a ms.; in German.   MSS 354 A

954. Govi, Gilberto, 1826-1890. *Letters.* 1859-1880. 2 items.

A.L.S. (1859 Jan. 28, Florence) to Moigno dealing with the polarization of the light of comets, and A.L.S. (1880 July 17, Naples) to Favaro concerning ancient compasses; in Italian.   MSS 607 A

955. Richardson, John, Sir, 1787-1865. *Letter.* 1859. 1 item (3 p.)

A.L.S. (1859 March 31, Lancrigg, Grasmere) to Owen thanking him for a copy of his lecture on the gorilla; Owen annotated the letter.   MSS 1209 A

956. Schönbein, Christian Friedrich, 1799-1868. *Letters.* 1859-1863. 4 items.

One A.L.S. (1859) to an unnamed correspondent, 2 A.L.S. and an A. envelope (1862-1863) to Nicklès; in German.   MSS 1335 A

957. Varley, Cromwell Fleetwood, 1828-1883. *Letter.* 1859. 1 item. (4 p.)

A.L.S. (1859 Nov. 10, Lothbury, London) to F. J. Till on letterhead of The Electric & International Telegraph Co'y, explaining that Schneider's idea of a spiral conductor is not new and that his invention is not useful for a Second Atlantic Cable.   MSS 1501 A

958. Balfour, John Hutton 1808-1884. *Letter.* 1860. 1 item (4 p.)

A.L.S. (1860 June 15, Edinburgh) to Mr. Duckworth.

MSS 42 A

959. Bright, Charles Tilston, Sir, 1832-1888. *Letter.* 1860. 1 item (2 p.)

A.L.S. (1860 Nov. 28, London) to Bayley concerning traffic on the Hull and Doncaster line.   MSS 177 A

960. Dalafosse, Gabriel, 1796-1878. *Letter.* 1860. 1 item (2 p.)

A.L.S. (1860 Dec. 1, Paris) to Milne-Edwards concerning the replacing of a professor of mineralogy; in French.   MSS 419 A

961. Esmarch, Friedrich von, 1823-1908. *Papers.* 1860-1875. 2 items.

A.L.S. (1875 Mar. 15), and holograph inscription (unsigned and undated) on the detached cover of an offprint, 1860; in German.   MSS 489 A

962. Hue, Louis. *Letter.* [186-?]. 1 item (2 p.)

A.L.S. (Sept. 25) to Jerdan requesting tickets to Drury Lane.   MSS 729 A

963. Jolly, Philipp Johann Gustav von, 1809-1884. *Letter.* 1860. 1 item (1 p.)

A.L.S. (1860 Aug. 9, Munich); in German.

MSS 758 A

964. Kopp, Hermann Franz Morietz, 1817-1892. *Letters.* 1860-1879. 6 items.

Six A.L.S. to various correspondents including Lucae, Mayer, and Scheurer-Kestner; in German.   MSS 795 A

965. Pasteur, Louis, 1822-1895. *Drawings of flasks.* [between 1860 and 1895]. 1 item: col. ill.

Pencil and watercolor drawings on thin board of swan-necked flasks probably by Pasteur for his studies on spontaneous generation and fermentation in the 1860's. "A reduire" written in pencil in upper left margin. [XIV A]

MSS 1605 A

966. Spencer, Herbert, 1820-1903. *Letters.* 1860-1901. 10 items.

Eight A.L.S. to various correspondents including De Morgan, Sieveking, and Hooker and 2 L.S. (1860 June 5 and 1901 Nov. 14) written when Spencer was ill.

MSS 1397 A

967. Tchihatchef, Pierre de, 1808-1890. *Letter.* 1860. 1 item (1 leaf)

A.L.S. (1860 Dec. 22, Paris) to an unnamed "confrère" offering a photograph of himself; in French.

MSS 1448 A

968. Whitworth, Joseph, Sir, 1803-1887. *Letter.* 1860. 1 item (2 p.)

A.L.S. ([18]60 Mar. 10, The Firs near Manchester) declining an invitation and inviting the unnamed correspondent to view Whitworth's gun. MSS 1562 A

969. Bezold, Albert von, 1836-1868. *Letters.* 1861-1867. 2 items.

Two A.L.S. (1861 May 15 and 1867 August, Bath); in German. MSS 110 A

970. Dessaignes, Victor, 1800-1885. *Letter.* 1861. 1 item (1 p.)

A.L.S. (1861 Nov. 13, Vendôme) thanking the recipient (Milne-Edwards?) for his compliments and a copy of his discourse; in French. MSS 434 A

971. Diesterweg, Friedrich Adolph Wilhelm, 1790-1866. *Letter.* 1861. 1 item (1 p.)

A.L.S. (1861 May 19, Berlin) apologizing for not being able to meet his correspondent in Cothen because he is busy in Berlin with sessions of parliament; in German. MSS 441 A

972. Field, David Dudley, 1781-1867. *Letter.* 1861. 1 item (1 p.)

A.L.S. (1861 Sept. 19, Stockbridge) to an unnamed correspondent concerning the death of Field's wife. MSS 510 A

973. Shaw, Norton, d. 1868. *Letter.* 1861. 1 item (1 leaf)

A.L.S. (1861 May 13) to an unnamed correspondent on letterhead of the Royal Geographical Society concerning a consultation "on the subject of the instruments." MSS 1363 A

974. Sheldrake, Thomas B. *[Practical notebook].* 1861-1862. [153] p.: ill. (one col.); 35 cm.

Bound holograph ms. signed twice (1861), detailing the construction and use of batteries and "apparatus for showing electrical phenomenon" and the procedures for electroplating, plaster casting, etc. Fly leaf signed and dated 1861 and "Exhibition 1862" cited near end of codex. Spine title: Electr[ical] Experiments. Six items laid in: an award from the Industrial Exhibition, Richmond (1881); an A.L.S. (1872 Nov.) to the Countess R.; and 4 sets of notes, one with the dates 1867-1870, another on letterhead of "Bot. of Thomas Sheldrake, Working Cabinet Maker." MSS 1279 B

975. Welcker, Friedrich Gottlieb, 1784-1868. *Letter.* 1861. 1 item (3 p.)

A.L.S. ([18]61 Mar. 23) to an unnamed friend; in German. MSS 1548 A

976. Welles, Gideon, 1802-1878. *Letter.* 1861. 1 item (2 p.)

A.L.S. (1861 Oct. 8, Navy Department) to Lewis & Goodwin concerning their "proposed improvement in telegraphic communications." MSS 1551 A

977. Challis, James, 1803-1882. *Letters.* 1862-1863. 2 items.

Two A.L.S., one (1862 Jan. 25, Cambridge) to Chambers, commenting on "A Handbook of descriptive and practical astronomy," and one (1863 Aug. 21, Cambridge) on zodiacal light. MSS 324 A

978. Fowler, John, Sir, 1817-1898. *Letter.* 1862. 1 item (2 p.)

A.L.S. (1862 June 3, Westminster) to Owen concerning the British Museum. MSS 533 A

979. Joule, James Prescott, 1818-1889. *Letters.* 1862-1867. 2 items.

A.L.S. (1862 Dec. 25, Manchester) supposedly to Séguin concerning researches in the dynamical theory of heat and the neglect by their contemporaries of important scientists, and A.L.S. (1867 Mar. 4, Manchester) declining the post of juror of the Imperial Commission at the Paris Exhibition. MSS 761 A

980. Reichenbach, Karl Ludwig Friedrich, Freiherr von, 1788-1869. *Papers.* 1862-1867. 162 [i.e., 184] leaves, [4], [1], [7]p.: port.; 33 cm.

Four physically separate holograph mss. signed, including "Physiologisch-pathologische Untersuchungen über Sensitivitaet und Somnambulismus," "Arwa, von [--?--] Bose" (1862 Sept.), "Analysen von [Arwa--?]" (1862 Oct. 7-1867 June 19); in German. With photographic portrait. MSS 1263 B

981. Silliman, Benjamin, 1816-1885. *Letters.* 1862-1870. 2 items.

A.L.S. (1862 Jan. 16, New Haven) to Gould concerning the proofs of "Tryon's sketch of the history of American conchology," and A.L.S. ([18]70 March 28, New Haven) on dropping "Jr." from his name. MSS 1374 A

982. Vincent, Alexandre Joseph Hidulphe, 1797-1868. *Essai de restitution de la chirobaliste d'Heron d'Alexandre, disciple de Ctésibius d'après les manuscrits de la bibliothèque impériale et d'autres documents.* 1862. [1], 23 leaves [i.e., 42 p.], [1] leaf: col. ill.; 35 cm.

French translation, dedicated to Napoleon III, of work supposedly by Hero of Alexandria. MSS 271 B

983. Armstrong, William George Armstrong, Sir, baron, 1810-1900. *Letters.* 1863-1871. 2 items.

A.L.S. (1871 July 6) from Armstrong to Cooke requesting Cooke's vote for Armstrong's friend Stuart Rendel at the Atheneum, and holograph signed (?) "Dine" thanking W. G. Simpson for a proposal of which Armstrong has no need, with envelope postmarked 1863 Aug. 20. MSS 21 A

984. Arnott, Neil, 1788-1874. *Letter.* 1863. 1 item (1 p.)

A.L.S. (1863 March 28) to J. D. Sowerby, Secretary of the Royal Botanic Society and Gardens, Regent's Park (1833-1869) apologizing for his absence at a visit by the Royal Family. MSS 22 A

985. Buckland, Francis Trevelyan, 1826-1880. *Letters.* 1863-1875. 6 items.

Six A.L.S. to various correspondents including Flower, one (1870 Mar. 14) introducing Lamont. Additional material includes newspaper clippings. MSS 199 A

986. Dammer, Otto, 1839-1916. *Papers.* 1863-1914. 3 items.

Two holograph draft letters concerning the *Handbuch der anorganischen chemie,* and a holograph outline of the *Handbuch;* in German. Additional materials include a booklet with Dammer's biography and a letter from Bruno Dammer. MSS 396 A

987. Figuier, Louis, 1819-1894. *Letters.* 1863. 3 items.

Three A.L.S.; in French. MSS 511 A

988. Geikie, Archibald, 1835-1924. *Papers.* 1863-1915. 6 items.

Five A.L.S., and autograph signature. MSS 581 A

989. Lindshelm, J. G. *Letter.* 1863. 1 item (1 leaf)

A.L.S. (1863 April 25, Washington) to Miss Keluits (?) concerning the taking of her photograph. MSS 901 A

990. Mansel, Henry Longueville, 1820-1871. *Letter.* 1863. 1 item (4 p.).

A.L.S. (1863 Jan. 31, Oxford) to an unnamed correspondent concerning miracles, especially the miracles of Christ. MSS 1606 A

991. Sainte-Claire Deville, Henri Etienne, 1818-1881. *Correspondence.* 1863-1878. 19 items.

Eighteen A.L.S. (eleven on letterhead of the Laboratoire de Chimie, Ecole Normale Supérieure, Paris), 10 of which are to "maitre," perhaps Chasles, and one A.L.S. (1863 April 29, Paris) to Sainte-Claire Deville from C. [Barosé?]; in French. MSS 1320 A

992. Avebury, John Lubbock, baron, 1834-1913. *Letters.* 1864-1895. 30 items.

Twenty-three A.L.S., 3 A. postcards S., and 4 L.S. to various correspondents including Browning, Flower, Herschel, Hooker, Huggins, Weismann, and Wilde. MSS 943 A

993. Bessemer, Henry, Sir, 1813-1898. *Letters.* 1864-1895. 4 items.

Four A.L.S. to various correspondents including Ashworth, Horsfield and Bragge. Additional material includes newspaper obituaries.　　MSS 107 A

994. Falconer, Hugh, 1808-1865. *Letter.* 1864. 1 item (4 p.)

A.L.S. (1864 Oct. 25, Montauban) to Sharpey presenting Falconer's reasons for seconding Darwin for the Copley medal.　　MSS 499 A

995. Hall, Granville Stanley, 1844-1924. *Letter.* [between 1864 and 1924]. 1 item (1 p.)

A.L.S. (Dec. 17, Brindisi) to an unnamed correspondent concerning his just-received note.　　MSS 655 A

996. Jones, John Winter, 1805-1881. *Letter.* 1869. 1 item (2 p.)

A.L.S. (1869 April 26, British Museum) to Dyce Duckworth accepting a dinner invitation from the Edinburgh University Club.　　MSS 759 A

997. Lesseps, Ferdinand de, 1805-1894. *Papers.* 1864-1888. 11 items.

Eight A.L.S., including one to Dupin; 2 autograph cards; 1 calling card with holograph note; in French.　　MSS 424 A

998. Reimer, Dietrich, 1818-1899. *Letter.* 1864. 2 items.

A.L.S. ([18]64 Feb. 18, Berlin) to Professor Dr. Barth, mentioning Leipzig, with envelope; in German.　　MSS 1201 A

999. Schlagintweit, Robert von, 1833-1885. *Letters.* 1864-1872. 2 items.

A.L.S. (1864 Mar. 21, Giessen) to an unnamed correspondent, and A.L.S. (1872 Aug. 20, Giessen in Hessen) to Herr Müller mentioning Wilson; in German.　MSS 1331 A

1000. Weierstrass, Karl Theodor Wilhelm, 1815-1897. *Papers.* 1864-1888. 5 items.

Two A.L.S., 2 long A.L.S. (1864, 1876) containing equations, and a notarized A.D.S. (1885) concerning the studies of Mathias Lerch; in German.　　MSS 1542 A

1001. Wilson, Daniel, Sir, 1816-1892. *Letter.* 1864. 1 item (3 p.)

A.L.S. (1864 Jan. 31, University College, Toronto) to Evans inquiring about "the flint implements found in the Cave-deposits."　　MSS 1573 A

1002. Wurtz, Charles Adolphe, 1817-1884. *Papers.* 1864-1871. 4 items.

Two A.L.S. (1864 Apr. 27 and 1871 July 20) on letterhead of the Faculté de Médecine de Paris, the latter to Darmstaedter, holograph note on a business card, and a holograph leaf of memorial sentiments concerning Laurent and Gerhardt (on cardboard); in French.

MSS 1589 A

1003. Bernard, Claude, 1813-1878. *Note.* [1870?]. 1 item (1 p.)

Holograph note beginning: Il y a deux foies. . .

MSS 93 A

1004. Cannizzaro, Stanislao, 1826-1910. *Letters.* 1865-1894. 2 items; port.

Two A.L.S. (1865 Sept. 12, Palermo and 1894 Jan. 8, Rome); in Italian.　　MSS 297 A

1005. Fizeau, Armand Hippolyte Louis, 1819-1896. *Papers.* 1865-1896. 12 items: ill.

Nine A.L.S., mostly to Sénarmont; A. note; holograph article "Cuivre oxydulé - premières experiences, 1865" (17 p.); and holograph article "Affaire de la Bibliothèque de l'Observatore," including letters and notes (12 leaves); in French.　　MSS 516 A

1006. Glaisher, James, 1809-1903. *Letters.* 1865-1870. 4 items.

Four A.L.S. to various correspondents including Hogg. Additional material includes newspaper obituary.

MSS 597 A

1007. Ladd, William, 1815-1885. *Letter.* 1865. 1 item (1 leaf)

A.L.S. (1865 Nov. 7, London) acknowledging a check.

MSS 810 A

1008. Mach, Ernst, 1838-1916. *Papers.* 1865-1918. 121 leaves: ill., ports. 34 cm.

Ninety-nine letters (1865-1915) and 40 postcards, typed and holograph. Correspondents include Popper-Lynkeus and Tausig. Seven calling cards, one telegram, and two holograph mss. Eight newspaper items, two memorial booklets (1916, 1918), and thirteen journal articles by Einstein, Gomperz, Popper, and others. Collected in leather-bound scrapbook; in German. Additional material includes Czechoslovakian and German seminar announcements.

MSS 909 B

1009. Mill, John Stuart, 1806-1873. *Letters.* 1865-1869. 3 items.

Three A.L.S. (1865 Nov. 5, 1868 Oct. 26, and 1869 Feb. 14) all from Avignon.

MSS 1005 A

1010. Noeggerath, Johann Jacob, 1788-1877. *Letter.* 1865. 1 item (1 leaf)

L.S. (1865 June 19, Bonn) to Ludwig; in German.

MSS 1077 A

1011. Riecke, Eduard, 1845-1915. *Note.* [between 1865 and 1915]. 1 item (1 leaf)

A. note S. (undated) concerning a table of contents, cancelled; in German.

MSS 1213 A

1012. Roscoe, Henry Enfield, Sir, 1833-1915. *Letters.* 1865-1891. 10 items.

Nine A.L.S. to various correspondants including Crompton, Meyer, and Scheurer Kestner; to the latter he wrote concerning Dr. Douglas Hogg who replaced Roscoe in a jury judging exhibits at the Exposition Universelle, Paris, 1889. Also, one A. card S. ([18]86 Aug. 5, Heidelberg) to Foster.

MSS 1231 A

1013. Sylvester, James Joseph, 1814-1897. *Letters.* 1865-1889. 2 items.

A.L.S. (1865 Aug. 30, [K. Woolwich?]) concerning the next meeting of the English Association, and A.L.S. (1889 April 25, New College, Oxford) to Jesse C. Green concerning Sylvester's scientific memoirs.

MSS 1441 A

1014. Tomlinson, Herbert, 1845-1931. *Letter.* [between 1865 and 1931]. 1 item (1 leaf)

A.L.S. (King's College, Strand, London) to an unnamed correspondent requesting copies of papers on physical science.

MSS 1477 A

1015. Tylor, Edward Burnett, Sir, 1832-1917. *Letter.* 1865. 1 item (3 p.)

A.L.S. ([18]65 Feb. 12, Wellington) to Evans concerning Evans' book on numismatics and Tylor's book on travels in Mexico.

MSS 1485 A

1016. Virchow, Rudolf Ludwig Karl, 1821-1902. *Letter.* 1865. 1 item (1 leaf)

A.L.S. (1865 June 14, Berlin); in German.

MSS 1509 A

1017. Blavier, Edouard Ernest, 1826-1887. *Letters.* 1866. 2 items.

Two A.L.S. (1866 July 11 and Aug. 14, Nancy); in French.

MSS 120 A

1018. Flammarion, Camille, 1842-1925. *Papers.* 1866-1907. 18 items.

Twelve A.L.S., to various correspondents including Young, 1 holograph page, 1 A. postcard S., 3 calling cards, and a holograph article "Comment sur la mesuré la terre"; in French.

MSS 517 A

1019. Hansteen, Christopher, 1784-1873. *Correspondence.* 1866. 2 items.

A.L.S. (1866 Sept. 27) in Danish, and holograph letter signed to Hansen (August 19, Copenhagen) in German.

MSS 661 A

1020. Barré de Saint-Venant, Adhémar Jean Claude, 1797-1886. *[Sur le choc longetudinal de deux barres parfaitement élastiques et sur la proportion de leur force vive qui est perdue pour la translation ulterieure].* 1867. 1 item (2 [i.e., 4] p.)

Holograph ms. S. of presentation to Société Philomattrique de Paris (1867 May 11); in French. MSS 1319 A

1021. Hudson, Robert, 1834-1898. *Letter.* 1867. 1 item (1 p.)

A.L.S. (1867 July 6, Clapham Common S.) to Jabez Hogg declining an invitation. MSS 728 A

1022. Keller, Ferdinand, 1800-1881. *Letter.* 1867. 1 item (3 p.): ill.

A.L.S. (1867 June 8) to Evans announcing the dispatch of 4 cases with antiquities from Swiss lake dwellings; in German. MSS 778 A

1023. Lankester, Edwin Ray, Sir, 1847-1929. *Letter.* [between 1867 and 1929]. 1 item (3 p.)

A.L.S. (May 9, South Kensington) to Lady Evans concerning the kindness of the Evanses during a vacation in Monte Carlo. MSS 822 A

1024. Lewis, William James, 1847-1926. *Letter.* [between 1867 and 1902]. 1 item (4 p.)

A.L.S. (undated) to Wiltshire concerning mineralogy. MSS 891 A

1025. Sorby, Henry Clifton, 1826-1908. *Letters.* 1867-1870. 2 items.

Two A.L.S. ([18]67 April 16, Hyde Park, [London] and [18]70 March 10, Sheffield) concerning microscopy. MSS 1392 A

1026. Cooper, Peter, 1791-1883. *Papers.* 1868-[1881?]. 4 items.

Two A.L.S. (1868 Jan. 27, [1881?] Oct. 25, New York), the former to Godwin concerning a wharf and pier system for New York, and two autograph cards. MSS 371 A

1027. La Villesboisnet, W. Espivent de. *Letters.* 1868. 2 items.

Two A.L.S. (1868 Mar. 18 and 24, London) to Moigno concerning an invention by Mr. Piggott; in French. MSS 857 A

1028. Percy, John, 1817-1889. *Letter.* 1868. 1 item (1 leaf)

A.L.S. (1868 Nov. 4) to Dr. Duckworth declining an invitation; on letterhead of the House of Commons. MSS 1113 A

1029. Stewart, Balfour, 1828-1887. *Letters.* 1868-1881. 2 items.

A.L.S. (1868 July 29) on letterhead of Kew Observatory, Richmond, Surrey, S.W., and A.L.S. (1881 Nov. 18) to Geo[rge] Yates on letterhead of The Owens College, Manchester. MSS 1420 A

1030. Suess, Eduard, 1831-1914. *Letters.* 1868-1909. 8 items: ill.

Three A.L.S. (1868-1894) in German to various correspondents including Redlich, A. envelope to Dr. Reuss, 3 A.L.S. (1909) in English to Prof. Sollas, and holograph note on a calling card; in German. MSS 1434 A

1031. Weld, Charles Richard, 1813-1869. *Letter.* 1868. 1 item (3 p.)

A.L.S. ([18]68 Dec. 3) on letterhead of the Bath Royal Literary & Scientific Institution, suggesting the use of common straw paper "as well as the more expensive stamped note paper." MSS 1549 A

1032. Babinet, Jacques, 1794-1872. *Letters.* [ca. 1869]. 3 items.

A.L.S. (undated) to unidentified M. le comte; A.L.S. (undated) to unidentified; A.L.S. (1869 Dec. 16) to a judge recommending Mme. Gervais's suit to him; in French. MSS 33 A

1033. Coles, Cowper Phipps, 1819-1870. *Letter.* 1869. 1 item (2 p.)

A.L.S. (1869 Nov. 12) to his daughter concerning her riding bodice and the health of relatives. MSS 362 A

1034. De La Rue, Warren, 1815-1889. *Letters.* 1869-1872. 2 items.

A.L.S. (1869 April 28) to Duckworth declining a speaking engagement, and A.L.S. (1872 Oct. 5, Cranford) to Hooker stating that Mr. Simmonds has withdrawn from the candidateship of librarian and remarking that they are photographing the scale on the Pagoda. MSS 421 A

1035. Derby, Edward George Geoffrey Smith Stanley, 14th earl of, 1799-1869. *Letter.* 1869. 1 item (5 p.)

A.L.S. (1869 June 24, London) to J. Taylor concerning the extension of the Alkali Act to cooper works. MSS 428 A

1036. Faye, Hervé Auguste Etienne Albans, 1814-1902. *Letter.* 1869. 1 item (4 p.)

A.L.S. (1869 Mar. 29) concerning meteors and shooting stars; in French.   MSS 503 A

1037. Frémy, Edmond, 1814-1894. *Letters.* 1869-1890. 4 items.

Four A.L.S.; in French.   MSS 544 A

1038. Hirzel, Heinrich, 1828-1908. *Letter.* 1869. 1 item (1 p.)

A.L.S. (1869 Nov. 28, Leipzig) concerning a prospectus on an "ice-cream-soda-apparatus"; in German.
                                                   MSS 705 A

1039. Jenner, William, Sir, 1815-1898. *Papers.* 1869-1889. 21 items.

Nineteen A.L.S. to various correspondents including Beresford-Hope and Sieveking, 1 autograph signature, and 1 envelope.   MSS 753 A

1040. Langley, Samuel Pierpont, 1834-1906. *Letters.* 1869-1887. 3 items.

Three A.L.S.   MSS 821 A

1041. Taylor, Alfred Swaine, 1806-1880. *Letter.* 1869. 1 item (1 leaf)

A.L.S. (1869 Sept. 9, Regents Park, [London]) to Miss Rosa Monckton concerning 3 photographs which he is sending her.   MSS 1446 A

1042. Bataille, Louis Nicolas, 1840-1896. *L'explication du fusil à aiguille faite par M. Louis Battaile.* [187-?]. 1 item (10 p.)   MSS 51 A

1043. *[Handbook of practical formulas].* [1874?]. [22], 180 p.: ill.; 21 cm.

A handbook containing various technical formulas, including astronomical and electrical formulas. Index at front.   MSS 264 B

1044. Hooker, Joseph Dalton, 1817-1911. *Correspondence.* 1870-1873. 25 items.

Twenty-five A.L.S. to Hooker from various correspondents including Balfour, Bentham, Carruthers, Farr, Gray, Miller, Percy, Prestwich, Ramsay, Rawlinson, Sclater, and Sharpey.   MSS 717 A

1045. Horsford, Eben Norton, 1818-1893. *Letter.* 1870. 1 item (2 p.)

A.L.S. (1870 Mar. 27, Cambridge) to an unnamed correspondent concerning the cost of printing a pamphlet.
                                                   MSS 723 A

1046. Moleschott, Jacob, 1822-1893. *Document.* 1870. 1 item (1 leaf)

A.D.S. (1870 May 30, Torino) with a prescription of some kind.   MSS 1020 A

1047. Popper-Lynkeus, Josef, 1838-1921. *Papers.* 1870-1917. 25 items: ill.

Sixteen A.L.S. (1870-1880) to various correspondents including Mach and Menger on various topics including drama, 4 A. postcards S. (1872-1873) with one on Kulke's play *Korah,* 1 A. envelope, 1 A. note S. (1917), A. ms. S.(?) (18 leaves) titled "Ernst Mach," A. ms. S. (4 leaves) titled "Zeppelin," and A. ms. (8 p.) titled "Über das Problem der Luftshiffahrt" with diagrams and a penciled note by the title: "An R. Mayer (1876)"; in German.
                                                   MSS 1161 A

1048. Richet, Charles Robert, 1850-1935. *Letters.* [between 1870 and 1935]. 5 items.

Four A.L.S. ([190]6, three undated) one recommending that the unnamed correspondent introduce Richet's student Michel Faquet to the bacterial biologist Urbain, and an A. card S. (undated) recommending Beaugrand to the Académie Française; in French.   MSS 1212 A

1049. Siemens, Carl Wilhelm, 1823-1883. *Letter.* 1870. 1 item (2 p.)

A.L.S. ([18]70 Dec. 24, Kensington Gardens) to Herr C. Koeltrer; in German.   MSS 1369 A

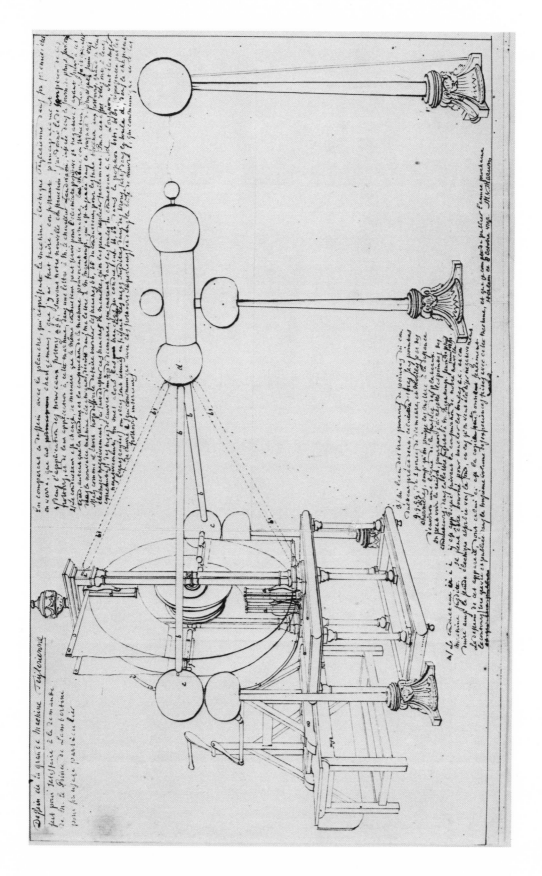

XI. A drawing for improving Teyler's static electricity machine (1792).

Entry No. 396, SI Neg. 84-7193.

1050. Strouhal, Čenek, 1850-1922. *Note.* [between 1870 and 1922]. 1 item (1 leaf)

Anonymous text on an envelope, headed "Affaire Weyr-Pexider," notes that Dr. Strouhal in the name of a mathematical society (?) was involved in correspondence with Weyr and Pexider; in German. MSS 1428 A

1051. Tedeschi, Vitale, 1854-1919. *[Applicazione della fotografia alla microscopia].* [1875?]. [7] p., [23] leaves of plates: chiefly ill.; 29 cm.

Holograph ms. signed, detailing modifications in photography to permit accurate, easily taken pictures through a microscope. The photographs include 3 of the equipment and 20 selected from Tedeschi's 160 photographs taken through a microscope; Spine bears date 1875; in Italian. MSS 1292 B

1052. Tweeddale, Arthur Hay, 9th marquis, 1824-1878. *Letter.* 1870. 1 item (4 p.)

A.L.S. (1870 Aug. 4, Chislehurst) to Mr. Flower concerning an inconvenient meeting. MSS 676 A

1053. Youmans, Edward Livingston, 1821-1887. *Letter.* [between 1870 and 1879]. 1 item (2 p.)

A.L.S. ([187-?], New York) to [Hult?] on letterhead of *Popular Science Monthly* concerning a conversation with William and Newgate about the journal *Mind* and concerning Youmans' failing eyesight. MSS 1591 A

1054. Berthelot, Marcellin Pierre Eugène, 1827-1907. *Papers.* 1871-1907. 12 items: port.

9 A.L.S. to various correspondents, including Matignon and Pingard and an envelope; a holograph sheet of notes; and an admission ticket to Berthelot's funeral; in French. MSS 100 A

1055. Cremona, Luigi, 1830-1903. *Drawing.* 1871. 1 item (1 p.)

Geometrical sketch (1871 Feb. 13) with note by F. Weyr explaining that the sketch was owned by his father, E. Weyr; in German. MSS 381 A

1056. Geikie, James, 1839-1915. *Letter.* 1871. 1 item (4 p.)

A.L.S. (1871 Dec. 5, Bathgate) to De Rance concerning a comparison of geological ages in various countries. MSS 582 A

1057. Helmholtz, Hermann Ludwig Ferdinand von, 1821-1894. *Papers.* 1871-1893. 12 items: port.

Six A.L.S., one to Dove, and one with envelope, 1 leaf (folded) of calculations, and 4 D.S., one to Carhart; in German. Additional material includes newspaper biographies. MSS 683 A

1058. Mendel, Gregor, 1822-1884. *Papers.* 1871-1882. 5 items: ill.

Two A.L.S., 2 D.S., and 1 A.S.; in German. MSS 991 A

1059. Nélaton, Auguste, 1807-1873. *Letter.* 1871. 1 item (1 leaf)

A.L.S. (1871 Jan. 1, Paris) recommending M. Anger (?) on letterhead of the Société aux Blessés des Armées de Terre et de Mer, Palais de l'Industrie; in French. MSS 1066 A

1060. Péligot, Eugène Melchior, 1811-1890. *Letter.* 1871. 1 item (1 leaf)

A.L.S. (1871 Jan. 31) of condolence on death of one Henri, identified in a penciled note as Viscomte [Rigneult?]; in French. MSS 1108 A

1061. Wundt, Wilhelm Max, 1832-1920. *Letter.* 1871. 1 item (3 p.)

A.L.S. (1871 Mar. 27) to an unnamed correspondent; in German. MSS 1588 A

1062. Bell, Alexander Graham, 1847-1922. *Letters.* 1872-1916. 11 items; ill.

Two A.L.S. (one to Watson with sketches); and 9 T.L.S. (one on geneology with handwritten marginalia). Additional material includes a letter (1905 Jan. 8) describing dinner at Bell's house signed E. E. H.; photocopies of the Bell letters; and a letter and exhibit on the telephone. MSS 86 A

1063. Dawson, John William, Sir, 1820-1899. *Letters.* 1872-1876. 2 items.

A.L.S. (1872 Dec. 13, Montreal) to Ramsay concerning glacially formed lakes, and A.L.S. (1876 April 24, Montreal) to Huxley asking for the return of Dawson's reptilian slides.   MSS 413 A

1064.   Gassiot, John Peter, 1797-1877. *Letter.* 1872. 1 item (4 p.)

A.L.S. (1872 Sept. 27) to Hooker concerning a sale of old wood and a gift of lily bulbs.   MSS 571 A

1065.   Gaudry, Albert, 1827-1908. *Letters.* 1872-1873. 3 items: port.

A.L.S. (1872 Oct. 1, Paris) to Evans with signed carte de visite photograph, thanking Evans for his hospitality during their recent trip to England, in French and English, and A.L.S. (1873 July 21) on letterhead of the Muséum d'historie naturelle, paleontologie, in French.   MSS 573 A

1066.   Huggins, William, Sir, 1824-1910. *Papers.* 1872-1902. 7 items.

Five A.L.S. to various correspondents including Thompson, and 2 D.S., one to Knight.   MSS 731 A

1067.   Liddell, Henry George, 1811-1898. *Document.* 1872. 1 item (1 leaf)

A.D.S. (1872 Nov. 27, Oxford), University matriculation form for Prince Leopold.   MSS 895 A

1068.   Abel, Frederick Augustus, Sir, 1826-1902. *Letter.* 1873. 1 item (3 p.)

A.L.S. (1873 February 28, Woolwich, Eng.) from Abel to [Hooker?].   MSS 3 A

1069.   Cayley, Arthur, 1821-1895. *Letter.* 1873. 1 item (4 p.)

A.L.S. (1873 March 19, Cambridge) concerning an algebraic formula.   MSS 316 A

1070.   Fergusson, William, Sir, 1808-1877. *Letter.* 1873. 1 item (1 p.)

A.L.S. (1873 Oct. 29) to W. B. Williams agreeing to be a patron for a dramatic entertainment on behalf of K. C. H.   MSS 505 A

1071.   Hopkinson, John, 1849-1898. *Letter.* 1873. 1 item. (3 p.)

A.L.S. (1873 April 10) to Mr. Cooke concerning a lampstand probably for a lighthouse lamp, on letterhead of the Lighthouse Department Glassworks near Birmingham.   MSS 721 A

1072.   Lorentz, Hendrik Antoon, 1853-1928. *Letters.* [between 1873 and 1928]. 2 items: port.

A.L.S. (1906 Nov. 10, Leiden) concerning 2 articles on electrons, and mentioning Levi-Civita, and A.L. signed with initials (undated) to Ehrenfest; in German.   MSS 922 A

1073.   Vernide de Corneillon, comtesse. *Letter.* 1873. 1 item (1 leaf)

A.L.S. (1873 Feb. 2) to Loumarin, Vaucluse, on letterhead embossed "VCG," requesting information on the Exposition de Vienne so that she can there represent her uncle, Philippe de Girard, "l'Inventeur de la filature mécanique du lin"; in French.   MSS 1507 A

1074.   Dreyer, John Louis Emil, 1852-1926. *Den personlige ligning ved astronomiske passage-observationer forsög til besvarelse af Kjöbenhavns Universitets Prisopgave i Astonomi for 1874.* 1874. 150 p.: folded ill.; 23 cm.

Gold medal essay, Univ. of Copenhagen, 1874; in Danish and English. Two A.D.S. concerning the essay competition laid in.   MSS 242 B

1075.   Ehrlich, Paul, 1854-1915. *Papers.* 1874-1915. 2 items; col. ill.

A.L.S. (Berlin), and corrected proof color plate I from Pinkus's *Lymphatische Leukämie* (1901?); in German. Additional material includes certificate of authenticity signed by Otto Adler.   MSS 475 A

1076.   Galton, Douglas, Sir, 1822-1899. *Letters.* 1874-1883. 3 items.

L.S. (1874 Jan. 27) to Barry concerning alterations of the bookcases for the Society of Antiquarians, and A.L.S. (1883 Dec. 8) and envelope to Mr. Eykyn.   MSS 566 A

1077. Mühlig, Johann Gottfried Gottlob, 1812-1884. *Letter.* 1874. 1 item (1 leaf)

A.L.S. ([18]74 Oct. 27, Frankfurt a. M.) to an unnamed correspondent; in German. MSS 1602 A

1078. Sewell, James Edwards, 1810-1903. *Letter.* 1874. 1 item (3 p.)

A.L.S. (1874 May 1, Oxford) apparently from Dr. Sewell, Warden of New College, treating in detail and transcribing an A.L.S. in Old French from William of Wykeham (1367 June) to John Lord Cobeham, concerning the ransom of the Duke of Bourbon. Additional material includes a facsimile (1858) of Wykeham's letter. MSS 1590 A

1079. Šimerka, Vaclav, 1819-1877. *Correspondence.* 1874. 1 item (2 p.)

A.L.S. (1874) from Mikeš to Šimerka with Šimerka's A.L.S. in response on verso, concerning mathematics; in Czech. MSS 1375 A

1080. Taylor, John Ellor, 1837-1895. *Letter.* 1874. 1 item (2 p.)

A.L.S. (1874 Nov. 11, Ipswich) to Mrs. Beaumont concerning pamphlets and "Mr. Mallett's paper on the 'Source of Volcanic Energy.'" MSS 1447 A

1081. Wallace, Alfred Russel, 1823-1913. *Letters.* 1874-1910. 22 items.

Sixteen A.L.S. (most from Dorset) to various correspondents including Besant, Carruthers (concerning orchids), Strang (concerning a portrait), Swinton, and Thompson (concerning electricity), 4 A. postcards S. (1889-1903) to various correspondents including Warming, Henry Howard, and Olaf Halvorsen (decrying Theosophy), 1 form letter completed and signed (1880), and 1 autograph signature. MSS 1526 A

1082. Brodie, Benjamin Collins, Sir, 1817-1880. *Letter.* 1875. 1 item (4 p.)

A.L.S. (1875 Nov. 3, Reigate) to Prof. Mill concerning copies of Brodie's articles in the Transactions. MSS 184 A

1083. Guthrie, Frederick, 1833-1886. *Letter.* 1875. 1 item (2 p.)

A.L.S. (1875 Nov. 4) to Mills concerning a list of articles written and published. MSS 635 A

1084. Pellegrini, Angelo. *Descrizione di tutti gli obelischi e colonne che trovarsi nelle piazze di Roma disposta in forma di guida da Angelo Pellegrini.* [last quarter of 19th century]. [226] p.: ill.; 22 cm.

History and description of the columns and obelisks in the plazas of Rome with transcriptions of their Latin, Greek, and Italian inscriptions. With five engraved plates of Trajan's Column, three other columns, and an obelisk; variously paginated; in Italian. Spine title: Obelischi. MSS 1250 B

1085. Tissandier, Gaston, 1843-1899. *Letters.* 1875-1884. 2 items.

A.L.S. (1875 Apr. 24, Paris) to an unnamed correspondent (who annotated the letter and the attached calling card) concerning the "catastrophe du *Zenith,*" and A.L.S. (1884 Mar. 31, Paris) to Helot on letterhead of *La Nature, Revue des Sciences illustrée;* in French. MSS 1470 A

1086. Todhunter, Isaac, 1820-1884. *Letter.* 1875. 1 item (2 p.)

A.L.S. (1875 Jan. 3, Redford Park) to "Bro. Silvanus" thanking him for his "very artistic card." MSS 1474 A

1087. Wichelhaus, Karl Hermann, 1842-1927. *Letter.* 1875. 2 items.

A.L.S. and A. envelope ([18]75 Apr. 2 and 3, Rome) to Salzmann concerning an event on 13 March and plans for 13 April; in German. MSS 1563 A

1088. Cesnola, Luigi Palma di, 1832-1904. *Letter.* 1876. 1 item (2 p.)

A.L.S. (1876 Dec. 19) to Mr. Hall concerning a dinner party. MSS 321 A

1089. Crookes, William, Sir, 1832-1919. *Papers.* 1876-1913. 18 items.

Ten A.L.S. and 6 T.L.S. to various correspondents including Besant, Carhart, Frazer, Jevons, Routledge,

Thompson, and Turner, an envelope, and a signature card including Bolton's and Carhart's signatures.

MSS 383 A

1090. Fresenius, Karl Remigius, 1818-1897. *Letter.* 1876. 1 item (1 p.)

A.L.S. (1876 Dec. 13, Wiesbaden) to Prof. Schersle on letterhead of the Chemisches Laboratorium; in German.

MSS 545 A

1091. Russell, Henry Chamberlaine, 1836-1907. *Letter.* 1876. 1 item (4 p.)

A.L.S. (1876 Apr. 8, Sydney Observatory) to an unnamed correspondent about the brightness of the stars of the southern and northern hemisphere.   MSS 1246 A

1092. Spottiswoode, William, 1825-1883. *Letter.* 1876. 1 item (1 leaf)

T.L.S. (1876 Nov. 22) with holograph postscript, to Pollock concerning a letter from Miss Barnard which concerns a statue of Faraday.   MSS 1400 A

1093. Tomlinson, Charles, 1808-1897. *Letter.* 1876. 1 item (4 p.)

A.L.S. ([18]76 Mar. 30, Highgate) to Prof. Mills concerning two sets of memoirs.   MSS 1476 A

1094. Witt, Otto Nicolaus, 1853-1915. *Letters.* 1876-1903. 4 items.

Two A.L.S. (1876 May 24 and 1877 Dec. 5, Brentford, Middlesex) to Meyer, one T.L.S. (1903 Apr. 21, Berlin) to A. Bosch on letterhead of the Organisations-Comité des V. Internationalen Congresses für angewandte Chemie, and holograph note on a calling card (1903 Mar. 24) to G. Krell; in German.   MSS 1580 A

1095. Hoff, Jacobus Henricus Van't, 1852-1911. *Papers.* 1877-1902. 8 items.

Four A.L.S., three A. postcards S., one to Meyer, one A. note S.; in German.   MSS 708 A

1096. Kohlrausch, Friedrich Wilhelm Georg, 1840-1910. *Letters.* 1877-1901. 5 items.

Three A.L.S., one possibly to Thompson, T.L.S. and L.S. to Carhart; all in German.   MSS 793 A

1097. Proctor, Richard Anthony, 1837-1888. *Letter.* 1877. 1 item (2 p.)

A.L.S. ([18]77 Sept. 15, Brighton) to E. Smith requesting the address of Dr. Lawson Tait.   MSS 1175 A

1098. Stoney, George Johnstone, 1826-1911. *Letters.* 1877-1908. 3 items.

A.L.S. (1877 June 30) to Foster concerning a radiometer and "Crookes's layer" in liquids, A.L.S. (1904 Mar. 8) to Crookes concerning the table of atomic weights, and A.L.S. (1908 Feb. 29) to Minchin inquiring about Jeans.   MSS 1423 A

1099. Walferdin, François Hippolyte, 1795-1880. *Letter.* [between 1877 and 1880]. 1 item (2 leaves)

A.L.S. (undated) with leaf laid in (Jan. 27) to Arago concerning astronomical observations made in 1877; in French.   MSS 1524 A

1100. Borchardt, Carl Wilhelm, 1817-1880. *Letters.* 1878-1880. 5 items.

Five A.L.S. to Hirschfeld in German.   MSS 142 A

1101. Carlleyle, A. C. L. *Letter.* 1878. 1 item (219 p.); 22 cm.

A.L.S. (1878 June 14, Shahpur Padrauna) to [J. H. Rivett-Carnac?] concerning the languages of India.   MSS 237 B

1102. Gegenbaur, Carl, 1826-1903. *Letter.* 1878. 1 item (1 p.)

A.L.S. (1878 Mar. 1, Heidelberg); in German.   MSS 579 A

1103. Gill, David, Sir, 1843-1914. *Letter.* 1878. 1 item (4 p.)

A.L.S. (1878 Oct. 10, Kensington) declining a dinner invitation for several reasons among which are the papers he is preparing on observations of Mars, and on Egypt.   MSS 593 A

1104. Haeckel, Ernst Heinrich Philipp August, 1834-1919. *Papers.* 1878-1918. 28 items: ill., port.

Ten A.L.S. in German and English, 5 A. postcards S., 5 autograph signatures, 1 postcard S. and 1 blank picture postcard, 4 envelopes, 1 portrait, and holograph ms. (35 p.) "Die heutige Entwickelungslehre in Verhältnisse zur Gesammtwissenschaft," in German. Recipients include Devrient, Lees, and Wollweber.   MSS 647 A

1105. Hatson, James C. *Letters.* 1878. 2 items.

A.L.S. (1878 July 22, Ann Arbor) concerning contributions to a statue of Le Verrier, and A.L.S. (1878 Aug. 14, Ann Arbor) reporting the discovery of a planet Le Verrier had predicted, both to Fizeau.   MSS 671 A

1106. Stearns, J. B. *Letter.* 1878. 1 item (6 leaves)

A.L.S. (1878 Aug. 17, Valentia) to W. H. Preece concerning instruments to measure the moon's physical attraction and Edison's experiments with electricity; written on 6 telegraph blanks.   MSS 1408 A

1107. Tollens, Bernhard Christian Gottfried, 1841-1918. *Biographical sketch.* [between 1878 and 1900?]. 1 item (1 leaf)

Manuscript leaf: containing a biographical sketch of Tollens; in German.   MSS 1475 A

1108. Agassiz, Alexander, 1835-1910. *Letters.* 1879-1903. 5 items: ill.

4 A.L.S. (one in French), one referring to a "Belknap machine," and one A. card S. Additional material includes photocopies of two letters.   MSS 66 A

1109. Blake, Clarence John, 1843-1919. *Letter.* 1879. 1 item (4 p.)

A.L.S. (1879 May 29, Boston) to Wolf concerning otology. Additional material in folder.   MSS 118 A

1110. Carlleyle, A. C. L. *Letter.* 1879. 84 p.; 20 cm.

A.L.S. (1879 April 23, near Chapra) to Rivett-Carnac concerning various anthropological questions referring to India.   MSS 236 B

1111. Dana, Edward Salisbury, 1849-1935. *Letter.* 1879. 1 item (4 p.)

A.L.S. (1879 May 2, New Haven) to an unnamed correspondent concerning his assistance in the analysis of a murder by arsenic poisoning.   MSS 397 A

1112. Edison, Thomas Alva, 1847-1931. *Papers.* 1879-1924. 6 items.

Three A.L.S., 2 T.L.S., and holograph paper concerning heating metal by electric current.   MSS 470 A

1113. Edison, Thomas Alva, 1847-1931. *Scrapbook.* 1879-1897. [34] p.: ill., ports.; 31 cm.

A.L.S. (1880 March 1, Menlo Park, N.J.) of thanks for Phalometer calculations; printed facsimile of letter to the editor, *Scribner's Monthly;* biographical extract from unidentified journal, 1879; and "The Life story of Edison" by Sarah A. Temple, published in *The Temple Magazine,* v. 1, no. 12. Cover title: Autos. &c. / Thomas Alva Edison.   MSS 251 B

1114. Klein, Felix, 1849-1925. *Letter.* 1879. 1 item (1 leaf)

A.L.S. (1879 Feb. 9, Munich); in German.   MSS 787 A

1115. Lauer, Gustav Adolph, 1808-1889. *Paper.* 1879. 1 item (1 leaf)

Holograph ms. (1879 Feb. 27, Berlin); in German.   MSS 850 A

1116. Naquet, Alfred Joseph, 1834-1916. *Letter.* 1879. 1 item (4 p.)

A.L.S. (1879 April 27, Brussels) to an unidentified correspondent concerning a conference.   MSS 1057 A

1117. Newcomb, Simon, 1835-1909. *Papers.* 1879-1882. 3 items.

A.S. and two A.L.S. on letterhead of the Nautical Almanac Office, Navy Department, Washington, D.C. One letter (1879 Jan. 8) is to Lovering, Chairman of the Rumford Committee, soliciting funds for an experiment on the speed of light. The other (1882 May 21) is to Dr. Peters.   MSS 1069 A

1118. Perkin, William Henry, Sir, 1838-1907. *Letters.* 1897-1905. 2 items.

A.L.S. (1897 Apr. 27) to Thompson, and an A.L.S. (1905 June 19) accepting an invitation. Both from The Chestnuts, Sudbury, Harrow.   MSS 1126 A

1119. Sande Bakhuyzen, Hendricus Gerardus van de, 1838-1923. *Letter.* 1879. 1 item (1 p.)

A.L.S. (1879 April 22) to unidentified concerning observations of the southern stars; in Dutch.   MSS 40 A

1120. Thorpe, Thomas Edward, Sir, 1845-1925. *Letters.* 1879-1912. 5 items.

Four A.L.S. (1879-1891) to Meyer, and 1 A.L.S. (1912 Feb. 18) to Thompson.   MSS 1466 A

1121. Brassai, Samuel von, 1800-1897. *Autobiography.* 1880. 1 item (4 p.)

Holograph ms. autobiography signed (1880 Sept. 22); in German.   MSS 173 A

1122. *Le Divagazioni d'un ignorante.* [188-?]. 124 p.; 31 cm.   MSS 256 B

1123. Finsen, Niels R., 1860-1904. *Note.* [between 1880 and 1904]. 1 item (1 p.)

A. note of thanks on a calling card; in Danish.   MSS 512 A

1124. Hughes, David Edward, 1831-1900. *Letters.* 1880-1897. 3 items.

Three A.L.S. to various correspondents including Thompson.   MSS 732 A

1125. Kekulé, August, 1829-1896. *Papers.* 1880-1887. 3 items.

Two A.L.S. (1880 Jan. 23, and 1887 Jan. 17, Bonn), and an A. list S.; in German.   MSS 777 A

1126. Lerch, Matyas, 1860-1922. *Papers.* [between 1880 and 1922]. 3 items.

A.L. probably to Hermite concerning politics and a formula on elliptical functions in French, and 2 scientific papers in Czech.   MSS 884 A

1127. Lommel, Eugen Cornelius Joseph von, 1837-1899. *Letter.* 1880. 1 item (2 p.)

A.L.S. (1880 Sept. 17, Erlangen) to Kentzler enclosing holograph biographical data; in German.   MSS 918 A

1128. Röntgen, Wilhelm Conrad, 1845-1923. *[Collection of materials by and about Röntgen].* 1880-1974. 136 items in one box; 37 cm.

A.L.S. (1890 Dec. 30, Würzburg), D.S. (1910 Sept. 15), 17 reprints of articles (1880-1891) by Röntgen, all in German, 20 reprints and 1 T. ms. (1896-1952) concerning Röntgen, 81 items of correspondence (1929-1974) concerning Röntgen with various correspondents including Michael Pupin, Otto Glasser, and Bern Dibner, with 15 sets of notes; some in German, mostly in English. Additional material includes an X-ray on glass (1897 Jan. 5) and several photographs of Röntgen, his laboratory equipment, medals commemorating him, and early diagnostic X-rays. [xv]   MSS 1269 B

1129. Smyth, Warington Wilkinson, Sir, 1817-1890. *Lectures on mining by Prof. Warington Smythe.* 1880-1882. [378] p.: ill. (some col.); 21 cm.

Holograph ms. signed on 3 pages by Brough, Smyth's student at the Royal School of Mines. Brough dates his lecture notes 1880-1881, a pasted-in book advertisement is dated 1882. Bound with a publication by Smyth entitled "A lecture on mining."   MSS 1282 B

1130. Freidel, Charles, 1832-1899. *Papers.* 1881-1886. 4 items; port.

A.L.S. (1881 June 16, Paris), 2 A. notes on calling cards, and 1 autograph signature; in French.

    MSS 549 A

1131. Greeley, Adolphus Washington, 1844-1945. *Letters.* 1881-1905. 3 items: col. map.

Two A.L.S. to Halsey, one with col. map of the Arctic Regions, and T.L.S. to Brainard.   MSS 617 A

1132. Haskins, C. C. *Letter.* 1881. 1 item (1 p.)

A.L.S. (1881 Sept. 5, Chicago) to Gaiffe concerning his letter to the Chicago Electrical Society, with interlinear French translation.   MSS 668 A

1133. Huggins, William, Sir, 1824-1910.
Note. [1881?]. 1 item (2 p.): ill.

A. note (June 7) concerning a paper on Uranus and Saturn read at the Royal Society, signed with initials W. H.　　　　　　　　　　　　　　MSS 700 A

1134. Kolbe, Hermann, 1818-1884. *Letter.*
1881. 1 item (2 p.)

A.L.S. (1881 Jan. 25, Marburg); in German.
　　　　　　　　　　　　　　　　　MSS 794 A

1135. Rounds, William M. F. *Letter.* 1881. 1
item (4 p.)

A.L.S. (1881 June 10, Nantucket, Mass.) to Lippincott, congratulating him on his recent marriage and thanking him for information on "the early history of the Anti-Slavery movement in this country."　　MSS 1238 A

1136. Schiaparelli, Giovanni Virginio, 1835-1910. *Letters.* 1881-1889. 3 items.

A.L.S. (1881 Mar. 18, Milan) and A.L.S. (Aug. 29) concerning scientific articles, both in Italian, and A. form S. (1889 June 17, Milan) on the history, equipment and personnel of the Royal Observatory of Milan, to Knight.
　　　　　　　　　　　　　　　　　MSS 1330 A

1137. Steinheil, Hugo Adolph, 1832-1893.
Autograph signature. 1881. 1 item (1 leaf)

Autograph signature and epigram (1881, Munich) on printed form of the Autographen-Album des Deutschen Reichs; in German.　　　MSS 1412 A

1138. Anderson, John, 1833-1900. *Letter.*
1882. 1 item (3 p.)

A.L.S. (1882 May 2, Calcutta) to [Flower?].
　　　　　　　　　　　　　　　　　MSS 79 A

**1139. Du Bois-Reymond, Emil Heinrich,
1818-1896.** *Die zweiundfünfzigste Versammlung der Britischen Naturforscher, Bericht von E. du Bois-Reymond.* 1882. 68 p.; 32 cm.

Holograph ms. of a Report on the 52nd meeting of the British Association for the Advancement of Science, Southampton, 1882. With a typed translation into English by I. H. Rosenfeld. Slipcase titled: Du Bois Reymond-Torbedinidae (1882).　　　MSS 250 B

1140. Nordenskiöld, Adolf Erik, 1832-1901.
Letters. 1882-1890. 2 items.

L.S. (1882 May 22, Stockholm) introducing Mlle. K. Pälman, in French, and A.L.S. (1890 Sept. 29, Stockholm) concerning the history of the cartography of the Pacific Ocean, in English.　　　　　　　MSS 1078 A

1141. Roberts, Frederick Sleigh Roberts, 1st earl, 1832-1914. *The Channel Tunnel.* 1882. 1 item (6 p.)

A. ms. S. (1882 Apr. 17) arguing against "the proposal to make a tunnel between Dover & Calais" because "a tunnel would be a source of great danger" as an invasion route.　　　　　　　　　　　　MSS 1221 A

1142. Bonney, Thomas George, 1833-1923.
Letters. 1883-1897. 2 items.

Two A.L.S., one (1883 Dec. 12) requesting the removal of Mr. J. Hawkshaw's name from the Erosion of Sea Coasts Committee; and one (1897 Nov. 6) congratulating Thompson on a speech.　　　MSS 141 A

1143. Evans, John, Sir, 1823-1908. *Letters.*
1883-1905. 3 items.

A.L.S. (1883 Mar. 14) concerning a Roman coin, A.L.S. (1888 Oct. 30) concerning the County Council, both to Mr. Fordham, and A.L.S. (1905 June 17) declining a dinner invitation from the Society of Chemical Industry.　　　　　　　　　　　MSS 492 A

1144. Fischer, Emil, 1852-1919. *Papers.*
1883-1912. 10 items.

Nine A.L.S., chiefly to Meyer, and a calling card with autograph signature; in German.　　MSS 513 A

1145. Fleischl von Marxow, Ernst von, 1846-1891. *Letter.* 1883. 1 item (3 p.)

A.L.S. (1883 Sept. 5, St. Gilgen) concerning the use of electricity in medicine; in German.　　MSS 519 A

1146. Herschel, Alexander Stewart, 1836-1907. *Letter.* 1883. 1 item (2 p.)

A.L.S. (1883 Sept. 1, Newcastle on Tyne) sending gifts and good wishes to Mr. and Mrs. Aldis for their expedition.　　　　　　　　　　　　MSS 692 A

1147. Hitchcock, Edward, 1828-1911. *Letters*. 1883. 2 items.

A.L.S. (1883 Mar. 16, Amherst) to Earle concerning a catalog of shells, and an A. list of geology books.

MSS 706 A

1148. Mach, Ernst, 1838-1916. *Papers*. 1883-1908. 17 items in 2 folders.

Three A.L.S., 3 A. postcards S., 5 T.L. signed with signature stamp, all to various correspondents including Rabel and Zeller, 2 calling cards, 1 D.S., A. ms. S. (18 p.), and 2 envelopes; in German. Additional material includes a copy of Rabel's article "Mach und die Realität der Aussenwelt." MSS 940 A

1149. Minkowski, Hermann, 1864-1909. *Letters*. 1883-1898. 11 items.

Eight A.L.S. in French, 1 A. postcard S. in German to Sommerfeld, and 2 A.L.S. to Minkowski from Frau Juste Minkowski. MSS 1013 A

1150. Mosetig-Moorhof, Albert, Ritter von, 1838-1907. *Letter*. 1883. 1 item (3 p.)

A.L.S. (1883 July 9, Vienna); in German.

MSS 1043 A

1151. Mundy, Jaromir, Freiherr von, 1822-1894. *Letter*. 1883. 1 item (1 leaf)

A.L.S. (1883 July 3, Vienna); in German.

MSS 1048 A

1152. Painlevé, Paul, 1863-1933. *Papers*. [between 1883 and 1933]. 2 items.

Holograph ms. (4 leaves), "La question minière," on the law of 1810, and A.L.S. (n.d. rue Séquier); in French. MSS 1123 A

1153. Preece, William Henry, Sir, 1834-1913. *Letters*. 1883-1913. 9 items.

Two A.L.S. to Carhart, one (1893) agreeing to write a conference paper on "Signalling through space by means of electromagnetic vibrations," 1 A.L.S. (1883) to Ansell, 1 A.L.S. (1895) to Thompson, 4 A.L.S. (all 1897) to P. V.

Luke, 2 about working with Marconi, and 1 T.L.S. (1913) to Carhart about the death of Kelvin. Additional material includes two newspaper clippings concerning Luke.

MSS 1170 A

1154. Rosenthal, Moriz, 1833-1889. *Letter*. 1883. 1 item (1 leaf)

A.L.S. (1883 Aug. 12, Vienna) to the Committee of the "Internation[alen] Elektrischen Ausstellung"; in German.

MSS 1232 A

1155. Shadbolt, George. *Letter*. 1883. 1 item (3 p.)

A.L.S. (1883 Jan. 19 and 22, Beechcroft, Camden Park, Chislehurst) to Jabez Hazy concerning the contributions of Shadbolt and Wenham to the development of the "glass parabolic condensor." MSS 1360 A

1156. Siemens, Werner von, 1816-1892. *Letters*. 1883-1890. 3 items.

Two A.L.S. ([18]83 Aug. 11 and [18]90 June 20, Berlin) to Köhler, and autograph signature; in German.

MSS 1370 A

1157. Smiles, Samuel, 1812-1904. *Letters*. 1883-1885. 2 items.

A.L.S. (1883 Feb. 8 and 1885 Feb. 23, London), the first to Mr. W. Clark, both concerning the preparation of books to be published by Mr. Nasmyth. MSS 1380 A

1158. Stein, Sigmund Theodor, 1840-1891. *Letters*. 1883. 2 items.

Two A.L.S. (1883 July 5 and 26, Frankfurt a[m] M [ain]) to the directors of the Comité der Internationalen Elektrischen Ausstellung concerning his "Elektrizität und Ner[-?-]enleben"; in German. MSS 1411 A

1159. Blumentritt, Ferdinand, 1853-1913. *Letters*. 1884-1887. 2 items.

Two A.L.S. (1884 June 27; 1887 March 7, Leitmeritz) to unidentified correspondents; in German. MSS 125 A

1160. Boltzmann, Ludwig, 1844-1906. *Papers*. 1884-1899. 4 items.

Three A.L.S. to various correspondents and a ms. note (1899 May 1) concerning a formula in mechanics, signed by Boltzmann and Mach; in German. MSS 136 A

XII. A letter of Michael Faraday to Richard Phillips (1831).

Entry No. 686, SI Neg. 85-9514.

1161. Daubrée, Auguste, 1814-1896. *Letter.* 1884. 1 item (3 p.)

A.L.S. (1884 June 2, Paris) to an unnamed correspondent on letterhead of the Ecole Nationale des Mines; in French. MSS 409 A

1162. Guillardeau. *Note sur l'electro-chimie médicale.* 1884. 1 item (15 p.)

Holograph ms. (1884 Mar. 28); in French.
MSS 633 A

1163. Kronecker, Leopold, 1823-1891. *Letter.* 1884. 1 item (1 leaf)

A.L.S. (1884 Feb. 26, Berlin); in German.
MSS 802 A

1164. Kropotkin, Petr Alekseevich, kniaz, 1842-1921. *Papers.* 1884-1897. 3 items.

Two A.L.S. and A. card S. MSS 803 A

1165. Lodge, Oliver Joseph, Sir, 1851-1940. *Papers.* 1884-1912. 16 items.

Twelve A.L.S. and one L.S. to various correspondents including Allen, Barrett, Flower, D'Arcy Thompson, and Silvanus Thompson, one autograph signature, and two envelopes. MSS 917 A

1166. Masaryk, Tomas Garrique, 1850-1937. *Letters.* 1884-1904. 2 items.

Two A.L.S. (1884 Feb. 2, and 1904 Jan. 14), the former with a typed addendum; in Czech. MSS 969 A

1167. Pender, John, Sir, 1816-1896. *Letter.* 1884. 1 item (2 p.)

L.S. (1884 Sept. 16, London) to B. B. Murray of the London Chamber of Commerce declining an invitation.
MSS 1112 A

1168. Poynting, John Henry, 1852-1914. *Letter.* 1884. 1 item (4 p.)

A.L.S. (1884 March 10, Edgbaston) to Thompson on the "fundamental proposition" and equations of Poynting's forthcoming paper "On the Transfer of Energy." MSS 1168 A

1169. Rayleigh, John William Strutt, baron, 1842-1919. *Correspondence.* 1884-1918. 14 items: ill.

Eight A.L.S. (1884-1902), 2 A. envelopes, 1 A.L.S. (1897) in reply from Rollo [Appleyard?], 2 A. postcards S. (1905-1918), the latter with T. address, and 1 autograph signature. Correspondents include Carhart, A. B. Cooper, Dr. Deetr, MacCarthy, and Thompson. Topics include electrochemistry. MSS 1429 A

1170. Searle, George Frederick Charles, 1864-1954. *Scientific notes.* [between 1884 and 1954]. 2 items.

Holograph ms. S. (1949 Nov. 9): "Extract from Maxwell's Paper, 'On the Mathematical Clarification of Physical Quantities,'" and holograph leaf of notes, each with drawings. MSS 1346 A

1171. Tisserand, François Félix, 1845-1896. *Letter.* 1884. 1 item (1 leaf)

A.L.S. (1884 Apr. 10, Paris) to Monsieur Maindron arranging for a messenger to deliver a ms.; in French.
MSS 1471 A

1172. Volhard, Jakob, 1834-1910. *Papers.* 1884-1908. 7 items.

Four A.L.S. (1884-1904, Halle), one to Nieszytka, two A. postcards S. (1908 Apr. 16 and May 8, Halle) to Speter, and 1 A. questionnaire S.; in German. MSS 1513 A

1173. Wohlgemuth, Emil von, b. 1843. *Letter.* 1884. 1 item (1 leaf)

A.L.S. (1884 July 17, Luxemburg) to an unnamed correspondent concerning Crown Prince Rudolf; in German.
MSS 1582 A

1174. Ditscheiner, Leander, 1839-1905. *Postcard.* 1885. 1 item (1 p.)

A. postcard S. (1885 May 18, Vienna) to Reuss; in German. MSS 445 A

1175. Hall, Asaph, 1829-1907. *Letter.* 1885. 1 item (2 p.)

A.L.S. (1885 July 1, Washington, D.C.) to Knight concerning Tycho Brahe's star being a comet.   MSS 768 A

1176. Hertz, Heinrich, 1857-1894. *Letters.* 1885-1893. 6 items.

Five A.L.S. and 1 A. postcard S.; in German.

MSS 697 A

1177. Kofoid, Charles Atwood, 1865-1947. *Document.* [between 1885 and 1947]. 1 item (1 p.)

A.D.S. (undated) concerning biology on letterhead of the autograph collection of Howes Norris, Jr.

MSS 792 A

1178. Kundt, August Adolph Eduard Eberhardt, 1838-1894. *Document.* 1885. 1 item (1 leaf)

A.D.S. (1885 Oct. 16, Strassburg) containing a classroom and laboratory schedule; in German.   MSS 806 A

1179. Planté, Gaston, 1834-1889. *Letter.* 1885. 1 item (1 leaf)

A.L.S. (1885 July 6, Paris) to a colleague praising his "Mémoire sur les batteries secondaires"; in French.

MSS 1145 A

1180. Radcliffe, Charles Bland, 1822-1889. *Papers.* 1885-1886. 6 items.

A.L.S. (1886 June 11, London) to an unnamed correspondent about measuring the heat of the earth, and 5 charts (each 1885 through 1886) of tides at the different phases of the moon.   MSS 1185 A

1181. Röntgen, Wilhelm Conrad, 1845-1923. *Letters.* 1885-1905. 3 items.

A.L.S. ([18]85 Nov. 17, Giessen) to W. Fischer, and A. card S. ([19]05 Apr. 21, Munich) with A. envelope, to Krauss; in German.   MSS 1228 A

1182. Stefan, Josef, 1835-1893. *Letter.* 1885. 1 item (1 leaf)

A.L.S. (1885 Oct. 27, Vienna) to an unnamed correspondent with the title "Excellenz"; in German.

MSS 1409 A

1183. Teall, Jethro Justinian Harris, 1849-1924. *Letter.* 1885. 1 item (6 p.)

A.L.S. (1885 Feb. 15, Kew, Surrey) to Sollas concerning laboratory apparatus.   MSS 1449 A

1184. Thomsen, Julius, 1826-1909. *Letter.* 1885. 1 item (1 leaf)

A.L.S. (1885 May 5, Copenhagen) to Meyer on letterhead of the Universitets Chemiske Laboratorium concerning his thermochemical research; in German.

MSS 1460 A

1185. Twain, Mark, 1835-1910. *Letter.* 1885. 1 item (2 p.)

A.L.S. (1885 Aug. 14, Elmira) concerning a telegraph to be recommended to the Baltimore and Ohio Telegraph Company.   MSS 355 A

1186. Weismann, August, 1834-1914. *Letter.* 1885. 1 item (3 p.)

A.L.S. (1885 Mar. 5, Freiburg) to an unnamed colleague; in German.   MSS 1544 A

1187. Arago, Etienne, 1802-1892. *Letter.* 1886. 1 item (3 p.)

A.L.S. (1886 March 26, Paris) from Arago to Jean François Gigoux concerning the placement of Gigoux's paintings in a new museum.   MSS 13 A

1188. Ball, Robert Stawell, Sir, 1840-1913. *Letters.* 1886-[1911?]. 7 items.

Seven A.L.S. to various correspondents chiefly concerning lecture tour arrangements.   MSS 43 A

1189. Boussinesq, Joseph, 1842-1929. *Letter.* 1886. 1 item (4 p.)

A.L.S. (1886 Mar. 5, Lille) commenting on an article by M. Calinon which Boussinesq's correspondent had sent; in French.   MSS 157 A

1190. Lister, Joseph Lister, baron, 1827-1912. *Letters.* 1886-1899. 3 items.

A.L.S. (1886 Oct. 10) to Buchanan concerning a dinner invitation, and A.L.S. (1899 Dec. 7) to Thompson concerning a suggestion to the Council, with envelope.

MSS 913 A

1191.   Morgan, Thomas Hunt, 1866-1945.
Autograph signature. [between 1886 and 1945]. 1
item (1 leaf)   MSS 1038 A

1192.   Ramsay, William, Sir, 1852-1916. *Letters.* 1886-1912. 7 items: ill., port.

Seven A.L.S. (1886-1912) primarily about research in chemistry, to various correspondents including Lupton and Thompson and perhaps Le Chatelier and Meyer; in French, German, and English.   MSS 1187 A

1193.   Winchell, Alexander, 1842-1891. *Letter.* 1886. 1 item (1 leaf)

A.L.S. (1886 June 7, Ann Arbor) to Carhart on letterhead of the University of Michigan urging him to accept a position at the University.   MSS 1574 A

1194.   Eiffel, Gustave, 1832-1923. *Letters.* 1887-1905. 2 items.

A.L.S. (1887 Feb. 13) stating plans to return to Brussels, and A.L.S. (1905 July 14, Sèvres) congratulating someone on nomination for some office; in French.
                    MSS 476 A

1195.   Fuchs, Immanuel Lazarus, 1833-1902. *Letter.* 1887. 1 item (3 p.)

A.L.S. (1887 July 25, Berlin); in German.
                    MSS 551 A

1196.   Mueller, Ferdinand Jacob Heinrich, Freiherr von, 1825-1896. *Letter.* 1887. 1 item (2 p.)

A.L.S. (1887 Mar. 25) to Owen introducing Forrest.
                    MSS 1045 A

1197.   Wiedemann, Gustav Heinrich, 1826-1899. *Letter.* 1887. 1 item (3 p.)

A.L.S. ([18]87 Jan. 22, Leipzig) to an unnamed friend; in German.   MSS 1564 A

1198.   Ingersoll, Robert Green, 1833-1899.
Papers. 1888-1896. 10 items: port.

Six A.L.S., 3 autograph signatures, and a photograph.
                    MSS 740 A

1199.   Kimball. *Electricity.* 1888-[189-?]. 5 v. (96 leaves each): ill.; 25 cm.

Lecture notes, taken by Charles Lane Parr. Vols. 1 and 2 labeled "Electricity I and II," Vol. 5 labeled "Light." Topics of Vol. 3 include the barometer, radiation, and capaillarity; those of Vol. 4 include the solar constant and selective absorption. With diagrams, tables, and charts. Vol. 1 dated 1888 Dec. on first page and 1888 Aug. on later page. Vol. 3 includes transcription of an essay on atoms by Clifford.   MSS 837 B

1200.   Miers, Henry Alexander, Sir, 1858-1942. *Letters.* 1888. 2 items.

Two A.L.S. (1888 Oct. 27 and 1888 Nov. 1) to an unnamed correspondent concerning the angle of crystals, on letterhead of the British Museum (Natural History).
                    MSS 1004 A

1201.   Poincaré, Henri, 1854-1912. *Letters.* 1888-1907. 11 items.

Nine A.L.S., most undated, including one (1892 June 20, Le Touquet) possibly to Appell about Darboux, several letters concerning affairs of the Académie des sciences, and 1 T.L.S. (1907 May 10, Paris) with A. envelope; in French.   MSS 1154 A

1202.   Wellner, Georg, 1846-1909. *Letter.* 1888. 1 item (1 leaf)

A.L.S. ([18]88 Oct. 11, Brünn) possibly to Popper-Lynkeus concerning Lippert's use of a card Wellner had signed; in German.   MSS 1552 A

1203.   Apostoli, Georges, 1847-1900. *Letter.* 1889. 1 item (2 p.)

A.L.S. (1889 Aug. 29, Paris) from Apostoli to unidentified concerning Apostoli's hearty support for the other's efforts.   MSS 11 A

1204. Freycinet, Charles [Louis de Saulses] de, 1828-1923. *Letter.* 1889. 1 item (2 p.)

A.L.S. (1889 Oct. 2) concerning the funeral arrangements for General Faidherbe, on letterhead of the Ministère de la Guerre; in French.   MSS 547 A

1205. Holden, Edward Singleton, 1846-1914. *Papers.* 1889-1895. 3 items.

Two A.L.S. to Knight, and A.D.S. (1889 May 30), i.e., an information sheet concerning equipment, staff, and history of Lick Observatory.   MSS 712 A

1206. Maxim, Hiram Percy, 1869-1936. *Autograph signature.* [between 1889 and 1936]. 1 item (1 leaf)   MSS 978 A

1207. Meldola, Raphael, 1849-1915. *Letter.* 1889. 1 item (2 p.)

A.L.S. (1889 Dec. 23) to Lascelles with corrections for Lascelles's book.   MSS 989 A

1208. Blümcke, Gustav Adolf, 1857-1914. *Papers.* 1890. 2 items.

A.L. to Blümcke, and a holograph ms. titled: "Bemerkung zu den Versuchen des Herrn Prof. Raoul Pictet über due Spannkrafts - Curve der 'Flüssigkeit Pictet'"; in German.   MSS 123 A

1209. Hempel, Walther Mathias, 1851-1916. *Papers.* 1890-1899. 2 items.

A.L.S. (1899 Jan. 6, Dresden), and autograph signature on the cover of his work "Gasanalytische methoden"; in German.   MSS 684 A

1210. Schimpf, Emil, b. 1857. *Vorläuflige instruction über zusammensetzung gebrauch des Luftschifferparks.* [1886?]. [228] p.

Notes from an unidentified manual on the construction and use of military airship parks. Signed at end: Schimpff, second Lieutenant im Badischen Pionier Bataillon Nr. 14.   MSS 68 B

1211. Arsonval, Arsène d', 1851-1940. *Letters.* 1891-1911. 6 items.

A.L.S. (undated) to M. le Senateur; A.L.S. (1892 July 12, St. Germain les Belles) to unidentified on a carte pneumatique; A.L.S. (undated) to unidentified; A.L.S. (1891 Feb. 3, St. Germain les Belles) to unidentified; A.L.S. (1911) to M. Savarit on a carte pneumatique; in French.   MSS 28 A

1212. Bogdanov, Anatolii Petrovich, 1834-1896. *Letters.* 1891. 2 items.

Two A.L.S. (1891 Dec. 30, and undated); in French and German.   MSS 132 A

1213. Harrison, Benjamin, 1833-1901. *Letter.* 1891. 3 items: ports.

L.S. (1891 Dec. 23, Executive Mansion) to the Congress transmitting a report on the electrification of the District of Columbia, with photographs of Harrison and his wife, Caroline Scott Harrison.   MSS 665 A

1214. Keeler, James Edward, 1857-1900. *Letters.* 1891-1900. 3 items.

Three A.L.S. (1891 Oct. 26, 1900 June 6, and 1900 July 21) to Knight.   MSS 774 A

1215. Milne, John, 1850-1913. *Letter.* 1891. 1 item (3 p.)

A.L.S. (1891 July 7, Newport, I. W.) to Sir Clements concerning the desirability of a seismic survey.   MSS 1011 A

1216. Scott, Edward John Long, 1840-1918. *Letter.* 1891. 1 item (2 p.)

A.L.S. (1891 Apr. 15) to an unidentified correspondent on letterhead of the British Museum, London, concerning "your label," which was treated with hydrosulfate and transcribed.   MSS 1343 A

1217. Alvan Clark & Sons. *Letter.* 1892. 1 item (1 p.)

T.L.S. (1892 July 25, Cambridgeport, Mass.) to Knight concerning the Bruce photographic telescope.   MSS 352 A

1218. Amundsen, Roald, 1872-1928. *Autograph signatures.* [early 20th century]. 3 items.

Three cards bearing autograph signatures, one a postcard (22 May 1912, Montevideo) reproducing a winter scene by M. Munk, printed in Vienna.   MSS 78 A

1219. Carhart, Henry Smith, 1844-1920. *Correspondence.* 1892-1915. 7 items.

Seven A.L.S. to Carhart from various correspondents; in English and German. Additional material includes 7 newspaper clippings on Carhart.　　　MSS 300 A

1220. Gray, Elisha, 1835-1901. *Papers.* 1892-1898. 5 items.

Three A.L.S., 2 to Carhart and Thompson, an envelope, and an autograph signature.　　MSS 615 A

1221. Meyer, Victor, 1848-1897. *Postcard.* 1892. 1 item (1 leaf)

A. postcard S. (1892 April 29, Heidelberg) to Frau A. Spies(?); in German.　　　MSS 1000 A

1222. Peary, Robert Edwin, 1856-1920. *Papers.* 1892-1916. 7 items.

Four A.L.S. (1892-1913), one to George E. Pond (1892), one to Osborne (1912); one T.L.S. (1909) to Stanley D. Gray; a sketch drawn and annotated by Peary (1893); and a T.L. (1916) to Peary with his initials and "O.K." Additional material includes dealer's description.　　　MSS 1109 A

1223. Siemens, Friedrich, 1826-1904. *Letter.* 1892. 1 item (1 leaf)

A.L.S. (1892 Mar. 28, Dresden) to Herr Gennet; in German.　　　MSS 1368 A

1224. Thompson, Silvanus Phillips, 1851-1916. *Papers.* 1892-1916. 7 items: ill.

Three A.L.S. (1892-1895) to Carhart concerning electricity, one A.L.S. (1895) to B. P. Lascelles, two A. cards S. (1910-1911) to Christy, and one A.L.S. (1916) from Thompson's widow to Carhart on a printed funeral remembrance.　　　MSS 1459 A

1225. Winslow, Lyttleton Stewart Forbes, 1844-1913. *Letter.* 1892. 1 item (3 p.)

A.L.S. (1892 March 26, Cavendish Square) to an unnamed correspondent concerning an interview of Winslow by the "Evening News" regarding influenza.　　　MSS 1576 A

1226. Ayrton, William Edward, 1847-1908. *Letters.* 1893-1894. 3 items.

A.L.S. (Nov. 22, no year) to Foster congratulating him on his election as President of the Society of Telegraph Engineers; A.L.S. (1893 May 27, London) to Carhart accepting an invitation to read a paper at the Electrical Congress at Chicago; A.L.S. (1894 Sept. 8, London) to Carhart concerning the international specifications of the Clark cell.　　　MSS 31 A

1227. Becquerel, Antoine Henri, 1852-1908. *Papers.* 1893. 2 items: port.

Latin motto with autograph, and A.L.S. (1893 Oct. 21) to M. Pingard concerning Marshal Macmahon's funeral; in French.　　　MSS 58 A

1228. Cantor, Moritz Benedikt, 1829-1920. *Letters.* 1893-1903. 4 items.

Two A.L.S. and 1 A. postcard S. (Heidelberg) with envelope addressed to Fink; in German.　　MSS 299 A

1229. Crossley, Edward, 1841-1904. *Letter.* 1893. 1 item (1 p.)

A.L.S. (1893 Nov. 10) to Knight stating that Crossley had offered his three-foot telescope to a prior correspondent.　　　MSS 385 A

1230. Du Bois, Henri Eduard Johan Godfried, 1863-1918. *Letters.* 1893-1901. 2 items.

A.L.S. (1893 July 29, Berlin) to Carhart declining a speaking invitation, and A.L.S. (1901 Dec. 31, The Hague) to Thompson concerning the Boer War and the study of magnetic molecules.　　MSS 454 A

1231. Euler-Chelpin, Hans Karl August Simon von, 1873-1964. *Papers.* [between 1893 and 1964]. 4 items.

Christmas card in French, with envelope (1934) in German, 2 autograph sheets, in German.　　MSS 491 A

1232. Ewing, James Alfred, Sir, 1855-1935. *Letters.* 1893-1899. 2 items.

A. postcard S. (1893 July 18, Cambridge) to Carhart concerning a paper for the Electrical Congress, and A.L.S. (1899 Feb. 18, Cambridge) to Thompson concerning the proposing of Edward Hopkinson's name for the Royal Society.　　　MSS 495 A

1233.   Ferraris, Galileo, 1847-1897. *Letter.* 1893. 1 item (2 p.)

A.L.S. (1893 July 19, Torino) to Pope concerning the shipment of apparatus to the Chicago World's Fair.

MSS 506 A

1234.   Forbes, George, 1849-1936. *Letter.* 1893. 1 item (1 p.)

A.L.S. (1893 July 23, Niagara Falls) to Carhart concerning arrival time at the Chicago Exhibition.

MSS 527 A

1235.   Galton, Francis, Sir, 1822-1911. *Letters.* 1893-1909. 7 items.

Seven A.L.S. to various correspondents including Evans and Stenhouse.   MSS 567 A

1236.   Glazebrook, Richard Tetley, Sir, 1854-1935. *Papers.* 1893. 2 items.

A.L.S. to Carhart concerning the copyright of a book, and an announcement of a meeting of the British Association for the Advancement of Science, with A.L.S. on it inviting Carhart to come to the meeting (1893 June 5, Nottingham).   MSS 598 A

1237.   Hospitalier, Edouard, 1852-1907. *Letter.* 1893. 1 item (1 p.)

A.L.S. (1893 June 20, Paris) to an unnamed correspondent concerning the subject of Hospitalier's speech to be delivered in New York and Chicago.   MSS 724 A

1238.   Mascart, Eleuthère Elie Nicolas, 1837-1908. *Letters.* 1893-1898. 2 items.

A.L.S. (1893 May 29, Paris) to Carhart declining an honor, and A.L.S. (1898 Nov. 3, Paris) to Thompson concerning Thompson's work on the electro-magnet; in French.   MSS 971 A

1239.   Mendenhall, Thomas Corwin, 1841-1924. *Letter.* 1893. 1 item (2 p.)

A.L.S. (1893 June 20) to Carhart concerning lecture plans on letterhead of the Office of the Superintendent, U. S. Coast and Geodetic Survey.   MSS 992 A

1240.   Pickering, Edward Charles, 1846-1919. *Letters.* 1893-1907. 2 items.

A.L.S. (1893 Oct. 16, Harvard College Observatory, Cambridge, Mass.) to Knight about the reasons why Harvard abandoned plans for an observing station in California, and T.L.S. (1907 Oct. 19, Cambridge, Mass.) to Knight about a reference book and the stars Alcor and Mizar.   MSS 1140 A

1241.   Remsen, Ira, 1846-1927. *Correspondence.* 1893-1904. 2 items.

T.L.S. (1893 May 20, Philadelphia) to Remsen from Cobb with A.L.S. (1893 May 24) to Cobb from Remsen on the bottom half of the page, and an A. note S. (1904 July 6) on letterhead of The Autograph Collection of Howes Norris, Jr.   MSS 1202 A

1242.   Turpin, Eugène, 1848-1927. *Letter.* 1893. 1 item (1 leaf)

A.L.S. (1893 July 5, Colombes) to M. E. Möser supplying the quotations he had requested; in French.

MSS 1484 A

1243.   Weyr, Emil, 1848-1894. *Papers.* 1893-1928. 3 items.

A.L.S. (1893 Mar. 1, Vienna) to an unnamed correspondent, holograph ms. S.: "Beiträge zur Curvenlehre" (142 p.), and A.L.S. ([19]28 Feb. 13, Brno) in Czech from his son [Frantisek?] Weyr confirming the handwriting; in German.   MSS 1557 A

1244.   Krogh, August, 1874-1949. *Autograph signature.* [between 1894 and 1949]. 1 item (1 leaf)

Autograph signature with envelope.   MSS 801 A

1245.   Preston, Thomas, 1860-1900. *Letter.* 1894. 1 item (4 p.)

A.L.S. (1894 July 15, Dublin) to Foster thanking him for both his list of errata for Preston's "book on Heat" and also his review of the book in *Nature.*   MSS 1171 A

1246.   Prestwich, Joseph, Sir, 1812-1896. *Letter.* [between 1832 and 1896]. 1 item (4 p.)

A.L.S. (June 19, Shoreham, near Sevenoaks) to Mr. Cooke concerning an invitation.   MSS 1172 A

1247. Sabatier, Paul, 1854-1941. *Papers.* 1894-1936. 7 items.

Five A.L.S. to colleagues, 1 leaf of a ms. signed, all in French, and 1 A. envelope to Dr. Armin Weiner, in German. MSS 1313 A

1248. Soxhlet, Franz, Ritter von, 1848-1921. *Letter.* 1894. 1 item (1 leaf)

A.L.S. (1894 Sept. 1, Packing) to the editors of *Nerresten Nachrichten;* in German. MSS 1396 A

1249. Swan, Joseph Wilson, 1828-1914. *Letters.* 1894-1897. 4 items.

Three A.L.S. to J. C. Green, Madame Sterling, and Thompson, and A. pass admitting bearer to Swan's lecture. MSS 1439.

1250. Carpenter, Henry Cort Harold, 1875-1940. *Abstract.* [between 1895 and 1940]. 1 item (1 p.)

Signed holograph memo titled "Abstract" concerning the development of crystals on a heated electro-deposited iron sheet. MSS 304 A

1251. Engler, Karl Oswald Viktor, 1842-1925. *Papers.* [ca. 1878]-1895. 2 items.

A.D.S. (ca. 1878) with autobiographical information on questionnaire, and A.L.S. (1895 Aug. 7); in German. MSS 553 A

1252. Hussey, William Joseph, 1862-1926. *Letter.* 1895. 1 item (2 p.)

A.L.S. (1895 Sept. 5, Palo Alto) to Knight concerning Swift's comet. MSS 737 A

1253. Liebermann, Carl Theodor, 1842-1914. *Letter.* 1895. 1 item (2 p.)

A.L.S. (1895 April 4, Berlin); in German. MSS 896 A

1254. Lowell, Percival, 1855-1916. *Papers.* 1895-1907. 4 items: ill.

A.L.S. (1895 Feb. 21, Boston), T.L.S. (1907 Dec. 17) to Knight with autograph postscript, envelope, and illustration, and T. note S. MSS 925 A

1255. Lummer, Otto, 1860-1925. *Letter.* 1895. 1 item (2 p.): ill.

A.L.S. (1895 Dec. 30) to Thompson concerning Lummer's mercury vapor lamp; in German. MSS 930 A

1256. Michelson, Albert Abraham, 1852-1931. *Letters.* 1895-1917. 2 items: port.

A.L.S. (1895 Mar. 21) to Carhart concerning a promising student, and A.L.S. (1917 Mar. 8, Chicago) to Mr. Bond concerning lecture tickets, both on letterhead of Ryerson Physical Laboratory, The University of Chicago. MSS 1003 A

1257. Rowland, Henry Augustus, 1848-1901. *Letters.* 1895. 2 items.

Two A.L.S. (1895 Jan. 21 and Feb. 7, Baltimore) to Carhart concerning "the specifications for the Clark cell" to be considered by the National Academy of Sciences, on letterhead of the Johns Hopkins University.

MSS 1241 A

1258. Tesla, Nikola, 1856-1943. *Letters.* 1895-1916. 8 items.

Two A.L.S., three T.L.S. (two to Carhart concerning an electrical cell), T.L.S. and T. envelope on illustrated company stationery (1916) to W. H. Reed concerning Tesla's ice machine, and an autograph signature.

MSS 1452 A

1259. Argyll, George John Douglas Campbell, 8th duke of, 1823-1900. *Letter.* 1896. 1 item (4 p.)

A.L.S. (1896 Feb. 12, Kensington) from Argyll to Prof. Heddle concerning authority of University Chancellor over an unnamed Museum and asking for a specimen of "[Walthamite?]" in connection with glacial theory.

MSS 16 A

1260. Cohn, Hermann Ludwig, 1838-1906. *Postcard.* 1896. 1 item (2 p.)

A. postcard S. (1896 March 2, Breslau) to Kirchhoff, Berlin; in German. MSS 359 A

1261. Drude, Paul Karl Ludwig, 1863-1906. *Letter.* 1896. 1 item (1 p.)

A.L.S. (1896 Nov. 23, Leipzig); referring to the *Jahrbuch der Physik;* in German.　　　　MSS 453 A

1262. Kearton, Richard, 1862-1928. *Letter.* 1896. 1 item (4 p.)

A.L.S. (1896 Mar. 19, Elstree, Hertfordshire) to an unnamed correspondent concerning Kearton's book on birds' nests and the St. Kilda wren.　　　MSS 773 A

1263. Kukis, R. *Letter.* 1896. 1 item (2 p.)

L.S. (1896 July 16, Tokyo) thanking H. A. Spaulding for a gift of electrical apparatus from the General Electric Company.　　　　MSS 804 A

1264. Nobel, Alfred Bernhard, 1833-1896. *Letter.* 1896. 1 item (2 p.)

A.L.S. (1896 Feb. 18, San Remo) to Ljungström; in Swedish.　　　　MSS 1075 A

1265. Sachs, Julius von, 1832-1897. *Letter.* 1896. 1 item (1 leaf)

A.L.S. ([189]6 Sept. 6, Brandenburg); in German.　　　　MSS 1316 A

1266. Barham, George, Sir, 1836-1913. *Letter.* 1897. 1 item (11 p.)

A.L.S. (1897 June) to W. E. A. Martin and the editor of *Science Gossip* concerning the geology of the northeast Kent coast.　　　MSS 47 A

1267. Bell, Robert, 1841-1917. *Letter.* 1897. 1 item (2 p.)

A.L.S. (1897 Dec. 6, Ottawa) to Frost thanking him for his letter. Additional material includes newspaper clipping on Bell.　　　MSS 88 A

1268. Bose, Jagadis Chunder, Sir, 1858-1937. *Letter.* 1897. 1 item (1 p.)

A.L.S. (1897 Jan. 6, Maida Hill) to Thompson concerning polarization of electric and Röntgen rays by the crystal epidote.　　　MSS 149 A

1269. Bramwell, Frederick Joseph, Sir, 1818-1903. *Letter.* 1897. 1 item (2 p.)

A.L.S. (1897 Jan. 13, Westminster) congratulating Thompson on the brillance of Thompson's Christmas lectures at the Royal Institution.　　　MSS 165 A

1270. Cohn, Ferdinand Julius, 1828-1898. *Letters.* 1897. 2 items.

A.L.S. (1897 Jan. 22, Breslau) of condolences to Storch, and A. note S. (1897 Nov., Breslau); in German.　　　MSS 358 A

1271. Heaviside, Oliver, 1850-1925. *Papers.* 1897-1919. 3 items.

A. postcard S. (1913 April 19, Torquay) to J. F. J. Bethenod concerning the mailing of proofs, 2 proof copies of 1 page of an article "Mathematics and the age of the earth," by Perry with holograph annotations, and a galley proof of an article "Attenuation of Electric Waves along Wires and their Reflection at the Oscillator," by Barton with holograph notes on the verso (2 leaves).　　　MSS 677 A

1272. Hull, Edward, 1829-1917. *Letter.* 1897. 1 item (4 p.)

A.L.S. (1897 Sept. 30) to De Rance concerning a joint geological report.　　　MSS 733 A

1273. Maxim, Hiram Stevens, Sir, 1840-1916. *Letter.* 1897. 1 item (3 leaves)

T.L.S. (1897 Nov. 3, London) to Thompson concerning the production of diamonds.　　　MSS 979 A

1274. Nansen, Fridtjof, 1861-1930. *Autograph signature.* 1897. 2 items.

A.S. (1897 March) to F. B.-P. on a map of the polar regions, and a second map showing the route of the "Fram," with marginal notations probably not in Nansen's hand.　　　MSS 1056 A

1275. Planck, Max, 1858-1947. *Papers.* 1897-1930. 14 items.

Five A.L.S. (1897-1930); 2 A. postcards S., one to Ehrenhaft (1929) and one to Armin Weiner; 4 A. envelopes; 2 A. sets of notes, one headed "Erster Teil" (1915), the

XIII. A paper by Louis Pasteur "Sur le dimorphisme dans les substances actives" (1854).

Entry No. 916, SI Neg. 79-10401.

other dated 1920; and A. ms. S., "Ueber die Grenzschicht [verdünater?] Elektrolyte" (1929?); in German.

MSS 1144 A

1276. Righi, Augusto, 1850-1920. *Letters.* [between 1897 and 1920]. 3 items.

A.L.S. (1897 Feb. 4) to Thompson in Italian, and 2 A.L.S. (undated, but one cites an article written in 1900) in French, at least one, probably both to Thompson, all concerning both Marconi and electricity and all on letterhead of the Istituto di Fisica della R[eale] Università di Bologna. MSS 1215 A

1277. Brush, Charles Francis, 1849-1929. *Letter.* 1898. 1 item (1 p.)

A.L.S. (1898 March 17, Cleveland) to Carhart concerning Carhart's article in *The Electrical World.*

MSS 194 A

1278. Carnegie, Andrew, 1835-1919. *Papers.* 1898-1919. 5 items: ports.

One T.L.S. (1898), one A.L.S. (no date). One autograph sheet (1919), one signed photogravure postcard portrait, and one U. S. stamp with Carnegie's portrait.

MSS 301 A

1279. Hedin, Sven Anders, 1865-1952. *Letter.* 1898. 1 item (3 p.)

A.L.S. (1898 July 6, Stockholm) to Colles concerning Methuens' publication plans. MSS 678 A

1280. Herschell, Farrer, 1837-1899. *Autograph signature.* 1898. 1 item (1 p.)

Autograph signature (1898 Dec. 9, Quebec) on a card of the International Commission. MSS 694 A

1281. Hilbert, David, 1862-1943. *Letters.* 1898-1915. 2 items.

A.L.S. (1898 Dec. 31, Göttingen) probably to Courant concerning mechanics and the irreducibility of algebraic equations, and A.L.S. (1915 Feb. 3, Göttingen) to Courant concerning a lecture given by Debye; in German.

MSS 701 A

1282. Roberts-Austen, William Chandler, Sir, 1843-1902. *Letter.* 1898. 2 items.

A.L.S. ([18]98 April 12) with A. envelope to Thompson on "the art of 'bronzing' a copper so as to make it look like bronze." After detailing this "quaint" method, Roberts-Austen notes that he uses the "Japanese, or Wet Method"; on letterhead of the Royal Mint.

MSS 1222 A

1283. Roux, Wilhelm, 1850-1924. *Postcard.* 1898. 1 item (1 leaf)

A. postcard S. (1898 Jan. 7, Halle) to *Die Berliner wissenschaft Correspondenz;* in German. MSS 1240 A

1284. Russell, William James, 1830-1910. *Letter.* 1898. 1 item (3 p.)

A.L.S. (1898 Sept. 15, St. Ives, Ringwood, Hants.) to Thompson asking him to give the inaugural address at Bedford College. MSS 1247 A

1285. Stanley, Henry Morton, 1841-1904. *Letters.* 1898-1900. 2 items.

Two A.L.S. (1898 Mar. 5 and 1900 Feb. 20, Whitehall, [London]) to Alfred Hauson. MSS 1404 A

1286. Thomson, Elihu, 1853-1937. *Letters.* 1898-1926. 3 items.

A.L.S. (1898 June 22) to Dr. Silvanus [Thompson] thanking him for the copy of "Two Tracts" by Boyle, A.L.S. (1917 Nov. 27) to Mr. Hutchinson, and A.L.S. (1926 June 26) to Harry [Meyner?]. The letter to Hutchinson is stamped "Autograph Collection of Dr. Max Thorek, Chicago." MSS 1461 A

1287. Krehbiel, Henry Edward, 1854-1923. *Letter.* 1899. 1 item (6 leaves): ill.

A.L.S. (1899 Nov. 23, New York) to Mason asking a question about the primitive musical bow. MSS 799 A

1288. Lemström, Selim, 1838-1904. *Letter.* 1899. 1 item (4 p.)

A.L.S. (1899 June 15, Helsingfors) to Thompson concerning Lemström's influence machine. MSS 878 A

1289. Warburg, Emil Gabriel, 1846-1931. *Letter.* 1899. 2 items.

A.L.S. (1899 Feb. 9, Berlin, N.W.) and A. envelope to Dr. P. Grasnick; in German. MSS 1530 A

1290. Reynolds, Henry, b. 1873. *Anmeldung-Buch des stud[iums] Chem[ie] herrn Henry Reynolds aus Manchester.* 1899-1902. 22 p.; 20 cm.

Printed course diary of the Universität Göttingen for the 1890's used by Reynolds to note his courses for 1899 to 1902; in German. Cover: Henry Reynolds. Physiko-chemisches Institut Göttingen.　　　MSS 1267 B

1291. Rücker, Arthur William, Sir, 1848-1915. *Letter.* 1899. 1 item (4 p.)

A.L.S. (1899 Apr. 16, Gledhow Gardens) to Thompson thanking him for his assistance with "the magnetite" and summarizing Rücker's lecture on "Wilder's globe" and magnetism.　　　MSS 1242 A

1292. Ryan, Harris Joseph, 1866-1924. *Postcard.* 1899. 1 item (2 p.)

T. postcard S. (1899 June 1, Cornell University) to Hammer thanking him for a copy of his article and "the illustration of the Edison Invention."　　　MSS 1312 A

1293. Tammann, Gustav, 1861-1938. *Letter.* 1899. 1 item (1 leaf)

A.L.S. (1899 Dec. 1) to an unnamed colleague; in German.　　　MSS 1443 A

1294. Unwin, William Cawthorne, 1838-1933. *Letter.* 1899. 1 item (2 leaves)

A.L.S. ([18]99 Aug. 25, Kensington, W.) to Thompson thanking him for papers.　　　MSS 1489 A

1295. Weber, Heinrich Friedrich, 1843-1912. *Letters.* 1899. 2 items.

Two A.L.S. (1899 Oct. 17 and Nov. 26, Zurich) to a colleague on letterhead of the Physikalisches Institut des Eidgenöss Polytechnikums, concerning Weber's visit to Newcastle on Tyne and his colleague's research on the voltaic cell; in German.　　　MSS 1525 A

1296. Dewar, James, Sir, 1842-1923. *Letters.* 1900-1908. 5 items.

Four A.L.S. and 1 envelope to various correspondents including Dixon and Thompson.　　　MSS 436 A

1297. Harcourt, Augustus George Vernon, 1834-1919. *Letter.* 1900. 1 item (4 p.)

A.L.S. (1900 Nov. 5, Oxford) to Thompson concerning the electrification of Christ Church, Oxford.　　　MSS 663 A

1298. Jervis-Smith, Frederick John, 1848-1911. *Letter.* 1900. 1 item (3 p.)

A.L.S. (1900 Dec. 18, Oxford) to Thompson concerning people and events at Oxford.　　　MSS 755 A

1299. Job, Paul. *Letter.* [between 1900 and 1977]. 1 item (1 p.)

A.L.S. (Nov. 6, Paris) to an unnamed correspondent; in French.　　　MSS 756 A

1300. Lapworth, Charles, 1842-1920. *Letter.* 1900. 1 item (3 p.)

A.L.S. (1900 Aug. 22, Birmingham) to Dawkins concerning Dawkins's useful testimony before a commission.　　　MSS 824 A

1301. Lee, Sidney, Sir, 1859-1926. *Letter.* 1900. 1 item (2 p.)

A.L.S. (1900 July 6) to Thompson with thanks for sending Thompson's book on Gilbert of Colchester and article on Peter Short.　　　MSS 869 A

1302. Lenard, Philipp Eduard Anton, 1862-1947. *Letters.* 1900-1930. 2 items.

A.L.S. (1900 Oct. 30, Kiel) to Thompson concerning Lenard's work on electricity, in English, and A. postcard S. (1930 May 3, Heidelberg), in German.　　　MSS 879 A

1303. Pictet, Amé, 1857-1937. *Papers.* [between 1900 and 1937]. 2 items.

A.L.S. (1932 Aug. 8, Dully par Bursiriel) to a friend about a proposed visit, and a calling card with a brief message and initials; in French.　　　MSS 1141 A

1304. Seton, Ernest Thompson, 1860-1946. *Autograph signature.* 1900. 1 item (1 leaf)

Autograph signature (1900 [Mar.?] 8, New York) on lower half of last leaf of a friendly A.L.S.　　　MSS 1358 A

1305. Harker, Alfred, 1859-1939. *Letters.* 1901. 2 items.

Two A.L.S. (1901 Jan. 10, and 1901 Mar. 24, Cambridge) to Sollas concerning a new method of analysis for various rocks and minerals. MSS 664 A

1306. Mechnikov, Il'ya Il'ich, 1845-1916. *Letter.* 1901. 1 item (2 p.)

A.L.S. (1901 Dec. 30, Paris) to an unnamed correspondent with thanks for benevolence; in French. MSS 997 A

1307. Parsons, Charles Algernon, Sir, 1854-1931. *Papers.* 1901-1924. 5 items.

A.L.S. (1913 Mar. 12, London), A. postcard S. (1924 Oct. 7) to Werner Leschner, L.S. (1901 June 5), a letter signed by Agnes E. Parsons, and an envelope. MSS 1103 A

1308. Hrdlička, Aleš, 1869-1943. *Letter.* 1902. 1 item (1 p.)

A.L.S. (1902 Mar. 24, Mexico City) to Mason concerning an exchange of publications between the United States National Museum, Washington, D.C., and the American Museum of Natural History, New York. MSS 726 A

1309. Lassar, Oscar, 1849-1907. *Letter.* 1902. 1 item (1 leaf)

T.L.S. (1902 Sept. 4, Berlin); in German. MSS 844 A

1310. Levi-Civita, Tullio, 1873-1941. *Papers.* 1902-1912. 2 items.

A.L.S. (1902 Mar. 16) to Dell'Agnola, and D.S. (1912 Oct. 12) concerning the Università degli studi di Padova; in Italian. MSS 889 A

1311. Menshutkin, Nikolai Aleksandrovich, 1842-1907. *Letter.* 1902. 1 item (2 p.)

A.L.S. (1902 June 16/29, Karlsbad) to Vincent Iokayer in Russian and German. MSS 993 A

1312. Murray, James Augustus Henry, Sir, 1837-1915. *Letters.* 1902. 2 items.

A.L.S. (1902 Jan. 22, Oxford) to Thiselton-Dyer concerning the word "oleander," and A.L.S. (1902 July 3, Oxford) to Thompson concerning the word "dynamo." MSS 1052 A

1313. Voigt, Woldemar, 1850-1919. *Letter.* 1902. 1 item (3 p.)

A.L.S. (1902 Mar. 3, Göttingen) to Thompson concerning their researches and that of Brewster; in German. MSS 1512 A

1314. Weiss, Edmund, 1837-1917. *Letter.* 1902. 1 item (1 leaf)

A.L.S. (1902 Jan. 17) to Herr Schwadron; in German. MSS 1546 A

1315. Weyr, Eduard, 1852-1903. *Letters.* 1902-1903. 3 items.

Three A.L.S. to colleagues; in Czech. MSS 1556 A

1316. Zsigmondy, Richard Adolf, 1865-1929. *Letter.* 1902. 1 item (1 leaf)

A.L.S. ([19]02 Jan. 30, Jena) to Wangerin; in German. MSS 1597 A

1317. Branly, Edouard, 1844-1940. *Letters.* 1903-1915. 4 items: port.

Four A.L.S. to unidentified correspondents, settling accounts and arranging appointments; in French. MSS 171 A

1318. Calmette, Albert, 1863-1933. *Letters.* 1903-1907. 2 items.

Two A.L.S. (1903 March 8 and 1907 July 21, Lille), one mentioning tuberculosis; in French. MSS 292 A

1319. Campbell, William Wallace, 1862-1938. *Letter.* 1903. 1 item (1 p.)

T.L.S. (1903 April 6, San Francisco) to Knight concerning intended testing of atmospheric conditions in the Sierra Madre and San Jacinto Mountains. MSS 294 A

1320. Duhem, Pierre Maurice Marie, 1861-1916. *Letter.* 1903. 1 item (4 p.)

A.L.S. (1903 Mar. 8, Bordeaux) on letterhead of the Laboratoire de Physique Théorique, Université de Bordeaux, possibly to Perrier, recommending Wintrebert as a student; in French.   MSS 460 A

1321. Henry, Alfred Judson, 1858-1931. *Document.* 1903. 1 item (1 p.)

D.S. (1903 May 1, Washington, D.C.) notifying Brainard of his election to the National Geographic Society.
                                                          MSS 685 A

1322. Koch, Robert, 1843-1910. *Letter.* 1903. 1 item (1 p.)

A.L.S. (1903 Dec. 15, Bulaways, Rhodesia) to Wasserman with envelope; in German.   MSS 790 A

1323. Potier, Alfred, 1840-1905. *Document.* 1903. 1 item (1 leaf)

D.S. (1903 Feb. 17) for an examen generale of the Ecole Polytechnique signed by Potier as the Examiner; in French.   MSS 1162 A

1324. Roberts, Isaac, 1829-1904. *Letter.* 1903. 1 item (1 leaf)

A.L.S. (1903 Mar. 2, Starfield, Crowborough, Sussex) to B. R. Baumgardt giving him permission to make slides from plates in Roberts' article in *Knowledge.*
                                                          MSS 1223 A

1325. Darboux, Gaston, 1842-1917. *Letter.* 1904. 1 item (2 p.)

A.L.S. (1904 July 13, Paris) to an unnamed correspondent requesting a requisition, on letterhead of the Institut de France, Académie des sciences; in French.
                                                          MSS 400 A

1326. Dedekind, Richard, 1831-1916. *Letter.* 1904. 1 item (2 p.)

A.L.S. (1904 Sept. 26, Braunschweig) acknowledging receipt of the *Heidelberg-Testkatalog* and the *Memorial Diary for Mathematicians* in which his death date was given as 1899; in German.   MSS 415 A

1327. Hale, George Ellery, 1868-1938. *Papers.* 1904-1921. 3 items.

A.L.S. and T.L.S. to Knight, and a holograph page S. with a quoted passage from one of Hale's books.
                                                          MSS 653 A

1328. Janet, Pierre, 1859-1947. *Letter.* 1904. 1 item (2 p.)

A.L.S. (1904 Sept. 26, Washington, D.C.) to Mr. Baldwin; in French.   MSS 750 A

1329. Markwald, Willy, 1864-1953. *Letter.* 1904. 1 item (2 p.)

A.L.S. (1904 Jan. 12); in German.   MSS 958 A

1330. Metcalf, Victor Howard, 1853-1936. *Letter.* 1904. 1 item (2 p.)

T.L.S. (1904 July 12, Washington, D.C.) to Carhart with thanks and reminiscences, on letterhead of the Office of the Secretary, Department of Commerce and Labor.
                                                          MSS 996 A

1331. Meyerhof, Otto, 1884-1951. *Letter.* [between 1904 and 1941]. 1 item (2 p.)

A.L.S. (June 16, Philadelphia) to Herr Goldstein, on letterhead of the Dept. of Physiological Chemistry, The School of Medicine, University of Pennsylvania; in German.   MSS 1001 A

1332. Schwarzschild, Karl, 1873-1916. *Letter.* 1904. 1 item (2 p.)

A. card S. (1904 Dec. 10, Göttingen) to L. D.; in German.   MSS 1342 A

1333. Scott, Robert Falcon, 1868-1912. *Letter.* 1904. 1 item (1 leaf)

A.L.S. (1904 Oct. 13, Antarctic Expedition) to Lady Evans on letterhead with the address 80 Royal Hospital Road, Chelsea, S.W., [London], accepting a luncheon invitation.   MSS 1344 A

1334. Spiegler, Edward, 1860-1908. *Note.* 1904. 1 item (1 leaf)

A. note on a calling card ([19]04 July 25, Altaussee) to a colleague concerning a promotion; in German.
                                                          MSS 1398 A

1335. Stark, Johannes, 1874-1957. *Letters.* 1904-1933. 12 items.

Six A.L.S. (1904-1920), 3 A. postcards S. (1918-1920), 1 engraved invitation with autograph address (1919), 1 T.L.S. (1922), and 1 T. postcard S. (1933), all to Fraulein Doktor Rabel; in German. MSS 1406 A

1336. Stevenson, Thomas, Sir, 1838-1908. *Letter.* 1904. 1 item (1 leaf)

A.L.S. (1904 July 6, London) to R. J. Friswell on torn letterhead of [--?--] Laboratory, Guy's Hospital, thanking him for his letter. MSS 1419 A

1337. Baker, Benjamin, Sir, 1840-1907. *Letter.* 1905. 1 item (1 p.)

A.L.S. (1905 Dec. 13) from Baker to Armstrong giving D. Moody permission to examine the iron works at Charing Cross. MSS 39 A

1338. Frank, Adolph, 1834-1916. *Papers.* 1905. 3 items.

A.L.S. (1905 Nov. 5, Charlottenburg), holograph article, and envelope, all to Dr. Lowenfeld; in German. MSS 536 A

1339. Muséum, national d'histoire naturelle (France) Bibliothèque centrale. *Visite de sa majesté Charles Ier Roi de Portugal et des Algarves et de M. le Président de la Republique Française au Muséum National d'Histoire Naturelle le 24 Novembre 1905.* 1906. 50 items in one box; 34 cm.

Printed souvenir volume (1906), 14 A. mss. primarily of speeches by Becquerel, Curie (on radium), Lippmann (color photography), and Perrier but also including lists of objects shown to or offered by the king of Portugal, T. ms., 34 A.L.S.; in French. Additional material includes proofs of a book. MSS 1157 B

1340. Shapley, Harlow, 1885-1972. *Papers.* [between 1905 and 1972]. 2 items.

T. ms. S. (6 p., with holograph revision): "Thirty Deductions from a Glimmer of Star Light," and T.L.S. (1922 Nov. 20) to Mrs. Mary S. Brown on letterhead of the Harvard College Observatory. MSS 1361 A

1341. Hooker, John D., 1838-1911. *Letter.* 1906. 1 item (2 p.)

A.L.S. with envelope (1906 Sept. 22, Los Angeles) to Knight concerning the Mt. Wilson telescope. MSS 715 A

1342. Miller, Oscar von, 1855-1934. *Letters.* 1906-1933. 4 items.

Four T.L.S., two to Zimmermann; in German. MSS 1009 A

1343. Thomson, Joseph John, Sir, 1856-1940. *Letters.* 1906-1927. 2 items.

A. card S. (1906 Dec. 2, Cambridge) to Professor Thompson thanking him for his congratulations, and A.L.S. (1927 May 15, Cambridge) to Sir John declining his invitation to view "the Eclipse." MSS 1462 A

1344. Graebe, Carl, 1841-1927. *Postcard and portrait.* 1907. 2 items: port.

A. postcard S. (1907 May 28, Paris) to Fraulein Fleck, in German, and portrait. MSS 608 A

1345. Hartland, Edwin Sidney, 1848-1927. *Letter.* 1907. 1 item (2 p.)

A.L.S. (1907 Sept. 23, Highgarth, Gloucester) concerning the Jewish custom of the bridegroom breaking a glass. MSS 667 A

1346. Lehmann, Otto, 1855-1922. *Letter.* 1907. 1 item (1 leaf)

A.L.S. (1907 March 7, Karlsruhe); in German. MSS 874 A

1347. Péladan, Joséphin, 1859-1918. *Letter.* 1907. 1 item (1 leaf)

A.L.S. (1907 Jan. 23) to Monod on their misfortunes; in French. MSS 1119 A

1348. Ritchey, George Willis, 1864-1945. *Letter.* 1907. 1 item (1 leaf)

A.L.S. (1907 Feb. 8, Pasadena) to Knight concerning details of the construction of the 100-inch mirror for the Mount Wilson Observatory and reasons for the tarnishing of mirrors in telescopes, on letterhead of the Solar Observatory, Mt. Wilson, California. MSS 1216 A

1349. Abbott, Charles Greeley, 1872-1973. *Letter.* 1908. 1 item (1 p.)

A.L.S. (1908 May 25, Mt. Wilson, Calif.) from Abbott to Wm. H. Knight declining a speaking engagement.

MSS 2 A

1350. Osler, William, Sir, 1849-1919. *Letters.* 1908-1917. 2 items.

A.L.S. (1908 Mar. [7?]) to Thompson concerning early treatment of syphillis, and A.L.S. (1917 Mar. 1) to Mrs. Potter concerning a book-worm plate.   MSS 1092 A

1351. Pérez, Charles, 1873-1952. *Letter.* 1908. 1 item (2 p.)

A.L.S. (1908 Nov. 20, Bordeaux) about his professional aspirations; on letterhead of the Institut de Zoologie; in French.   MSS 1115 A

1352. Schlesinger, Ludwig, 1864-1933. *Papers.* 1908-1928. 5 items.

A.L.S. ([19]28 Sept. 22) mentioning a letter of Gauss (1807), 2 A. envelopes (one S.) both to Armin Weiner, and 2 holograph ms. S.: "Sur quelques problèmes paramétriques de la théorie des équations différentielles linéaires" (15 leaves, presented in 1908, published in 1909; in French) and "Über ein Problem der Diophantischen Analysis bei Fermat, Euler, Jacobi und Poincaré" (22 leaves, published in 1908). To each ms., Schlesinger added the publication information at the top; in German.

MSS 1333 A

1353. Einstein, Albert, 1879-1955. *Papers.* 1909-1948. 26 items: ill.

Four A.L.S., 1 A. postcard S., and 12 T.L.S. to various correspondents including Frank, Sommerfeld, Veblen, Wattenberg, Weinek, and White; 2 holograph sheets of calculations; 2 A.S., one with sketch; A. poem S.; T. ms. with galley sheets; 2 holograph ms. S. ("Anhang Eddington's Theorie" and "Über eine Ergazung"); in English and German. Additional material includes A.L.S. from Cardozo to Flexner (1932 Oct. 11). [XVI]   MSS 122 A

1354. Grenfell, Wilfred Thomason, Sir, 1865-1940. *Letter.* 1909. 1 item (1 p.)

T.L.S. (1909 March 25, St. Paul) to Sanger declining an invitation from the Historical Society, and requesting more slides from Labrador.   MSS 621 A

1355. Marconi, Guglielmo, marchese, 1874-1937. *Lecture delivered at the Royal Academy of Science, Stockholm, on 11th December 1909 on the occasion of the award to him of a Nobel Prize for Physics.* 1909. 45 leaves: ill.; 27 cm.

T.D.S. (1909 Dec. 10) on discoveries in telegraphy. Holograph annotations and emendations. Marconi cites several colleagues, notably Thomson and Fleming. Cover title in Swedish.   MSS 865 B

1356. Nicolardot, C. *Letter.* 1909. 1 item (3 p.)

A.L.S. (1909 Mar. 9, Paris) on polymers and other complex molecular combinations; on letterhead of the "Section technique del'Artillerie"; in French.

MSS 1073 A

1357. Pupin, Michael Idvorsky, 1858-1935. *Papers.* 1909-1929. 3 items.

Autograph signature with epigram (1909 April 23, New York), T.L.S. (1929 May 24, New York) to Herbert V. Prochnow, Chicago, and printed/typed contract (1922 Dec. 30) between Pupin and Charles Scribner's Sons to publish *From Immigrant to Inventor,* signed by Pupin and Scribner.   MSS 1180 A

1358. Shackleton, Ernest Henry, Sir, 1874-1922. *Letter.* 1909. 1 item (1 leaf)

A.L.S. (1909 Nov. 19, London) to [Grove-]Hills, Treasurer of the Royal Astrological Society, resigning his membership on letterhead of the British Antarctic Expedition 1907. The date the letter was received and acknowledged is noted at the top.   MSS 1359 A

1359. Welch, William Henry, 1850-1934. *Letter.* 1909. 1 item (1 leaf)

A.L.S. (1909 Mar. 18, Baltimore) to Norris on letterhead of The Autograph Collection of Howes Norris, Jr., containing a Latin "sentiment."   MSS 1547 A

1360. Wiener, Otto Heinrich, 1862-1927. *Letters.* 1909-1927. 31 items.

Fifteen T.L.S., 3 A.L.S., 8 A. postcards S., 1 T. post-card S., 2 A. cards S. (all from Leipzig) to Rabel, 1 holograph note on a calling card, and 1 A.L.S. ([19]27 May 7, Leipzig) to Rabel from L. Wiener (Wiener's widow?); in German.   MSS 1568 A

1361. Zeppelin, Ferdinand, Graf von, 1838-1917. *Letter.* 1909. 1 item (1 leaf): ill.

A.L.S. ([19]09 Mar. 24) on elaborate letterhead of Palast-Hotel, Berlin; in German.   MSS 1596 A

1362. Abraham, Max, 1875-1922. *Card.* [1910]. 1 item (2 p.)

Autograph card signed ([19]10 November 17, Milan, Italy) from Abraham supposedly to J. Laub, physicist, on the problems of Einstein's theory of relativity.
MSS 4 A

1363. Baeyer, Adolf von, 1835-1917. *Papers.* 1910. 2 items: port.

A.S. (1910 Nov. 15, Munich) with chemical symbol for indigo on letterhead of Howes Norris; A.L.S. (n.d.) from Baeyer to unidentified; in German.   MSS 35 A

1364. Brashear, John Alfred, 1840-1920. *Letter.* 1910. 1 item (1 p.)

T.L.S. (1910 Feb. 5, Pittsburg) to Knight concerning Brashear's catalog of astronomical and physical instruments.   MSS 172 A

1365. Buchner, Eduard, 1860-1917. *Letter.* 1910. 1 item (3 p.)

A.L.S. (1910 Nov. 3, Breslau); in German.
MSS 196 A

1366. Curie, Marie, 1867-1934. *Postcard.* 1910. 1 item (1 p.)

A. postcard S. (1910 Dec. 4, Paris) to Prof. Lakowitz regretting that she cannot attend a conference; in French. Additional material in folder.   MSS 387 A

1367. Neuberg, Carl, 1877-1956. *Papers.* 1910-1916. 3 items.

A. postcard S. (1910 Nov. 17, Charlottenburg) to Pohl, A. postcard S. (1916 Aug. 20, [Oberbauere?]) to Wiechowski, and A. calling card; in German.   MSS 1095 A

1368. Plate, Ludwig Hermann, 1862-1937. *Letters.* 1910. 2 items.

A.L.S. (1910 Mar. 10, Zoologisches Institut) on cremation, and A. postcard S. (1910 April 2, Zoologisches Institut, Jena) to Frau W. Storch Kuhlmann; in German.   MSS 1146 A

1369. Addison, Christopher, 1869-1951. *Letter.* 1911. 1 item (1 p.)

A.L.S. (1911 May 29, London) to Beddard recommending S. Hunter for some work at the zoo.
MSS 65 A

1370. Aitken, Robert Grant, 1864-1951. *Letter.* 1911. 1 item (1 p.)

T.L.S. (1911 Aug. 15, Mt. Hamilton, Ca.) to Knight concerning spectroscopic binaries.   MSS 62 A

1371. Boucart, Jacques, 1891-1965. *Letter.* [between 1911 and 1965]. 1 item (1 p.)

A.L.S. (Paris); in French.   MSS 153 A

1372. Dreyer, John Louis Emil, 1852-1926. *Notes for an edition of Brahe's works.* 1911-1926. 3 boxes, 1 bound vol. (46 p.)

Holograph ms. of Dreyer's edition of Tycho Brahe's complete works, 15 v., published 1913-1929; and a collation of Vienna mss. of Brahe's observations by H. Krumpholz; in English, Danish, Latin, and German. Boxes titled: Dreyer, Tycho Brahe, Observationes I-III; volume titled: Dreyer, Tycho Brahe, Observationes.   MSS 249 B

1373. Ehrenfest, Paul, 1880-1933. *Papers.* 1911-1925. 3 items.

A.L.S. (1911 Feb. 22) possibly to J. Laub, A. postcard S. (1918 Sept. 19) and holograph article "Energieschwankingen im Strahlungsfeld oder Krystallgitter bei Superposition quantisierter Eigenschwingungen," (22 leaves) printed in *Zeitscrift für Physik*, 1925; in German.   MSS 472 A

1374. Laub, Jacob, 1882-1962. *Correspondence.* 1911-1961. 54 items in 2 folders: ill.

Four T.L.S. (1954) to Dessauer, 1 typed scientific paper in draft with illustrations, 8 postcards and 38 letters from various correspondents including Berndt, Dietrich, Ebert, Ruechardt, Schädel, Sittkus, 2 miscellaneous papers, and 1 envelope; in German.   MSS 848 A

1375. Newton, Richard Bullen, b. 1854. *Letter.* 1911. 1 item (2 p.)

A.L.S. (1911 Nov. 25) to widow of Col. Beddome in sympathy. On letterhead of the British Museum (Natural History).   MSS 1071 A

1376. Svedberg, Theodor, 1884-1971. *Papers.* 1911-1934. 2 items.

Holograph ms. (5 leaves) (1911 May, Upsala): "Ultramicroskopische Beobachtung einer Temperaturkoagulation, von The. Svedburg und Katsuji Inouye," and A. envelope ([19]34 June 13) to Dr. Armin Weiner; in German.   MSS 1438 A

1377. Vanvelde, A. J. J. *Letter.* 1911. 1 item (2 p.)

A.L.S. (1911 Mar. 3, Gand) to an unnamed correspondent on letterhead of the Laboratoire Communal requesting that one of his assistants, M. [Honiewski?] be allowed to study at the Institut Pasteur during the summer; in French.   MSS 1499 A

1378. Arrhenius, Svante August, 1859-1927. *Correspondence.* 1912-1923. 8 items: port.

Four A.L.S. (1912 Oct. 24, 1923 June 28, 1923 July 26, 1923 Oct. 27) to unidentified; 1 T.L. to Pregl (1923 Nov. 16); 1 T.L. from Pregl to Arrhenius (1923 July 29); 1 T.L. possibly from Pregl to unidentified mentioning Arrhenius (1923 Nov. 30); and 1 T.L.S. from Pregl to unidentified mentioning Arrhenius, with handwritten correction (1923 Nov. 24); in German.   MSS 23 A

1379. Carrel, Alexis, 1873-1944. *Letter.* 1912. 1 item (2 p.)

A.L.S. (1912 Oct. 20, New York) to Mr. Downer declining an invitation to a celebration in Carrel's honor; in French.   MSS 306 A

1380. Perrier, Edmond, 1844-1921. *Letter.* 1912. 1 item (2 p.)

A.L.S. (1912 Oct. 12, Paris) on teaching geology; on letterhead of the Museum d'Histoire Naturelle; in French.   MSS 1116 A

1381. Schuster, Arthur, Sir, 1851-1934. *Letter.* 1912. 1 item (1 leaf)

A.L.S. (1912 July 2, Victoria Park, Manchester) to Thompson thanking him for his "beautiful translation of Huygen's classical book."   MSS 1341 A

1382. Ehrenfest, Tatiana (Afanassjewa), b. 1876. *Article.* 1913. 1 item (9 leaves)

Holograph article, "Zur frage über die Koncentrationsschwankungen in Radioactiven Losungen," published in in *Physikalische Zeitschrift,* 1913; in German. Additional material includes offprint.   MSS 473 A

1383. Geothals, George Washington, 1858-1928. *Papers.* 1913-1923. 4 items.

Two T.L.S., an A. postcard S., and a T.D.S.   MSS 602 A

1384. Hahn, Otto, 1879-1968. *Letters.* 1913-1967. 2 items: port.

A.L.S. (1913 Feb. 5, Berlin-Dahlem) in German with photograph attached, and T.L.S. (1967 Jan. 12, Göttingen) in English to Marshall Bean hoping he is not as ill as he fears and sending a reprint of some work by Hahn.   MSS 649 A

1385. Meyer, Edgar, b. 1879. *Letters.* 1913-1914. 2 items.

Two A.L.S. (1913 Nov. 24, Tübingen, and 1914 Jan. 13, Tübingen), the former to Laub; in German.   MSS 998 A

1386. *Nobel Prize Winners' signatures.* [after 1913]. 17 items.

Sixteen A.S. with short typed biographical sketches, and 1 addressed envelope.   MSS 1076 A

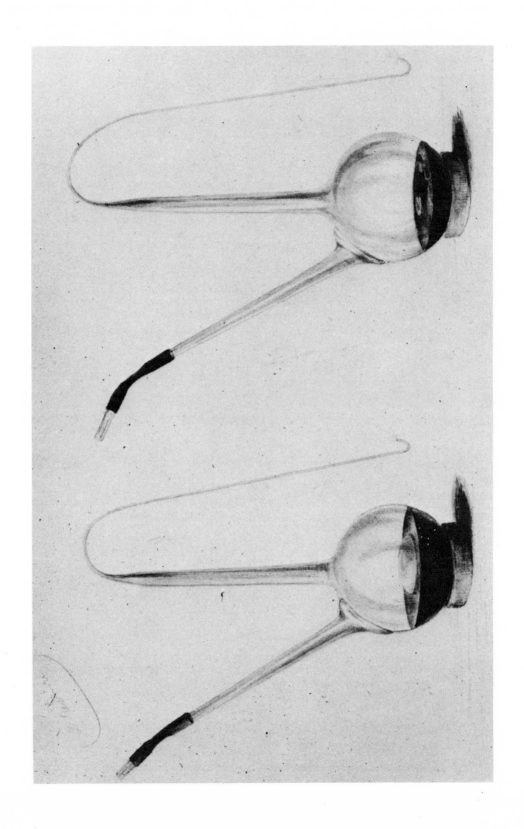

XIV A.   Drawings of flasks from the laboratory of Louis Pasteur
(ca. 1860).
Entry No. 965, SI Neg. 84-7174.

B. A certificate of vaccination signed by Louis Pasteur (1886).
Entry No. 916, SI Neg. 79-10409.

1387. Rutherford, Ernest, 1871-1937. *Letters.* 1913-1923. 2 items.

A.L.S. (1913 Feb. 1, Physical Laboratories, The University of Manchester) to Messrs. Sherratt & Hughes requesting that they stock Willows' *A text-book of physics,* and A. card S. ([19]23 Feb. 28, Cavendish Laboratory, Cambridge) to Pupin introducing J. A. Carroll. MSS 1248 A

1388. St. John, Charles Edward, 1857-1935. *Letter.* 1913. 1 item (1 leaf)

A.L.S. (1913 Feb. 13, Pasadena, California) to Knight on letterhead of the Mount Wilson Solar Observatory concerning St. John's recent lecture. MSS 1318 A

1389. Wien, Wilhelm, 1864-1928. *Letters.* 1913-1914. 2 items.

Two A.L.S. ([19]13 Dec. 2 and 1914 Sept. 22, Wurzburg, the latter on letterhead of the Physikalisches Institut der Universität) to an unnamed fraulein (perhaps Rabel); in German. MSS 1567 A

1390. Wright, Orville, 1871-1948. *Papers.* 1913-1930. 2 items.

T.L.S. (1913 June 19, Dayton, Ohio) to Zahm of the Smithsonian Institution on letterhead of The Wright Company regretting his inability to attend a meeting, and autograph signature (1930 June 12). MSS 1587 A

1391. Guggenheim, F. *Postcard.* 1914. 1 item (1 p.)

A. postcard S. (1914 Jan. 25, Berlin) to Pick; in German. MSS 631 A

1392. Rathenau, Walther, 1867-1922. *Letter.* 1914. 1 item (3 p.)

A.L.S. ([19]14 July 20, Schloss Freienwalde, Mark) to Blei; in German. MSS 1193 A

1393. Salkowski, Ernst Leopold, 1844-1923. *Letter.* 1914. 1 item (1 leaf)

A.L.S. ([19]14 Oct. 15, Charlottenberg, [Berlin]) to a colleague; in German. MSS 1322 A

1394. Lucas, Frederic Augustus, 1852-1929. *Letter.* 1915. 1 item (1 leaf)

A.L.S. (received 1915 Oct. 21) to Pratt concerning recent fires in Plymouth, on letterhead of the American Museum of Natural History, New York. MSS 929 A

1395. Willstätter, Richard Martin, 1872-1942. *Letters.* 1915-1932. 3 items.

Two A.L.S. (1926 Dec. 16 and 1932 Jan. 2, Munich) to unnamed colleagues (perhaps Ehrenhaft), and one A. card S. (1915 June 10, Berlin) to an unnamed doctor responding to his letter (of condolence?); in German. MSS 1572 A

1396. Ellerman, Ferdinand, 1869-1940. *Letter.* 1916. 1 item (1 p.)

A.L.S. (1916 May 18, Mt. Wilson) to Knight declining a speaking invitation. MSS 479 A

1397. Legouis, Emile Hyacinthe, 1861-1937. *Letter.* 1916. 1 item (3 p.)

A.L.S. (1916 April 10, Paris) to Baldwin concerning American neutrality; in French. MSS 872 A

1398. Quetelet, George. *Postcard.* 1916. 1 item (2 p.)

A. postcard S. (1916 Feb. 5, Brussels) to Weber about 7 letters (1821-1849) from Ørsted to Quetelet's relatives Adolphe Quetelet and Mons; in French. MSS 1183 A

1399. Walcott, Charles Doolittle, 1850-1927. *Letter.* 1916. 1 item (1 leaf)

T.L.S. (1916 Dec. 29, Washington, D. C.) to Woodhouse on letterhead of the Smithsonian Institution concerning dirigibles. MSS 1522 A

1400. Cottrell, Frederick Gardner, 1877-1948. *Letter.* 1917. 1 item (1 p.)

A.L.S. (1917 July 12, Washington, D. C.) to Howes Norris Jr.; in Esperanto. MSS 376 A

1401. Dyson, Frank Watson, 1868-1939. *Letter.* 1917. 1 item (1 p.)

A.L.S. (1917 Dec. 29, Greenwich) to Barley concerning understanding Newton's mind by means of the *Principia.* MSS 466 A

1402.　Einstein, Albert, 1879-1955. *Letters.*
1917-1941. 28 items.

One A.L.S. (1932 March 29), 18 T.L.S, handwritten answers to 7 typed queries, 1 T.L., and 1 telegram, all to Ehrenhaft, 1917-1941; in German. Additional materials include typed copies of Ehrenhaft's letters to Einstein, typescript of an article entitled "Meine Erlebnisse mit Einstein 1908-1940," letters from Rona, and papers concerning patents by Magnus; in German, English and French.　　　　　　　　　　　　MSS 289 B

1403.　Formanek, Emanuel, b. 1869. *Letter.*
1917. 1 item (1 p.)

A.L.S. (1917 Sept. 12); in German.　　MSS 528 A

1404.　Léon, Xavier, 1868-1935. *Letter.* 1917.
1 item (4 p.)

A.L.S. (1917 June 9, Paris) to Monsieur Baldwin on letterhead of the *Revue de Métaphysique et de Morale* concerning the *Revue,* the book *Children of Flame,* and American involvement in the war; in French.
　　　　　　　　　　　　　　　　MSS 1601 A

1405.　Picard, Emile, 1856-1941. *Letter.* 1917.
1 item (1 leaf)

A.L.S. (1917 Dec. 4, Paris) to a friend supplying a partial index to "mon discours sur Darboux"; on letterhead of the Académie des Sciences, Institut de France; in French.　　　　　　　　　MSS 1138 A

1406.　Bigourdan, Guillaume, 1851-1932. *Papers.* 1918. 2 items: ill.

Holograph letter to Gauthier-Villars (undated); galley sheet of an article by Marquet to be published in *Bulletin of Astronomy* (1918 Sept. 21) with Bigourdan's name in upper right-hand corner; in French.　　MSS 113 A

1407.　Laue, Max von, 1879-1960. *Papers.*
1918-1948. 12 items: ill.

Seven T.L.S. and three T. postcards S. to various correspondents including Fajans, Pelzer, Pringsheim, and Thirring, one typed envelope, and one T. scientific paper S. in draft "Ueber Heisenbergs Ungenauigkeitsbeziehungen und ihre erkenntnistheoretische Bedeutung"; in German.
　　　　　　　　　　　　　　　　MSS 849 A

1408.　Le Chatelier, Henri Louis, 1850-1936.
Letters. 1918-1927. 2 items.

T.L.S. (1918 May 27, Paris) to an unnamed correspondent convoking a meeting, and T.L.S. (1927 Oct. 8, Paris) to an unnnamed correspondent concerning reparation demanded by someone damaged by an article in the *Tribune de Paris;* in French.　　　MSS 863 A

1409.　Linde, Carl Paul Gottfried, Ritter von, 1842-1934. *Letters.* 1918-1927. 3 items.

A.L.S. (1918 Aug. 1, Munich) with envelope, and A.L.S. (1927 May 16, Munich); in German.
　　　　　　　　　　　　　　　　MSS 898 A

1410.　Michaelis, Leonor, 1875-1949. *Postcard.* 1918. 1 item (1 leaf)

A. postcard S. (1918 Jan. 17, Brandenburg) to Wiechowski; in German.　　　　　　MSS 1002 A

1411.　Osborn, Henry Fairfield, 1857-1935.
Letter. 1918. 1 item (1 leaf)

T.L.S. (1918 June 29) to Thayer concerning the Crocker Land Expedition and the eggs of the Knot.
　　　　　　　　　　　　　　　　MSS 1091 A

1412.　Pauli, Wolfgang Josef, 1869-1955.
Postcard. 1918. 1 item (1 leaf)

A. postcard S. (1918 Nov. 15, Vienna) to Wiechowski; in German.　　　　　　　　　MSS 1125 A

1413.　Berthelot,　Daniel,　1865-1927.
Letter. 1919. 1 item (2 p.)

A.L.S. (1919 Nov. 20, Paris) recommending M. J. Voicou; in French.　　　　　　　　　MSS 101 A

1414.　Caullery, Maurice Jules Gaston Corneille, 1868-1958. *Letter.* 1919. 1 item (1 p.)

A.L.S. (1919 Feb. 2) concerning a committee's deliberations; in French.　　　　　　　　MSS 315 A

1415.　Haber, Fritz, 1868-1934. *Papers.* 1919.
12 items: ill.

A. postcard S. (1919 July 28) to Wirth in German, 1 picture postcard with no message, and 10 photographs (6 different pictures).　　　　　　　MSS 645 A

1416. Langevin, Paul, 1872-1946. *Letter.* 1919. 1 item (2 p.)

A.L.S. (1919 Feb. 9, Paris) with a breakdown of costs for the Ecole Municipale de Physique et de Chimie Industrielles; in French. MSS 820 A

1417. Marcolongo, Roberto, 1862-1943. *Postcard.* 1919. 1 item (2 p.)

A. postcard S. (1919 Nov. 20, Naples) to Favaro; in Italian. MSS 959 A

1418. Schmidt, Ernst Albert, 1845-1921. *Postcards.* 1919. 3 items.

Three A. postcards S. (1919, Marburg), two to Wiechowski, one to Kransky(?); in German. With printed obituary notice containing photograph. MSS 1334 A

1419. Wieland, Heinrich, 1877-1957. *Papers.* 1919-1934. 7 items: ill.

Four A.L.S. (Munich), two on letterhead of the Organisch-chemisches Laboratorium der Kgl. technischen Hochschule, one holograph leaf signed with initials, one T. envelope to Armin Weiner, and one holograph ms. ([19]34 Aug. 20): "Die Konstitution von Bufotenin und Bufotenidin. Über Krötengiftstoffe VII; von Heinrich Wieland, Wilhelm [Kons?] und Heinz Mittant" (16, i.e., 18 leaves); in German. MSS 1566 A

1420. Cohn, Lassar, 1858-1922. *Letters.* 1920. 2 items.

Two A.L.S. (1920 Feb. 8 and Feb. 27, Köningsberg), both to Herr Elder, who was translating Cohn's "Arbeitsmethoden"; in German. MSS 845 A

1421. Debye, Peter Josef William, b. 1884. *Papers.* 1920-1929. 5 items.

T.L.S. with envelope, A. card S., T.L., and 2 holograph papers S.; in German. MSS 414 A

1422. Maxim, Hudson, 1853-1927. *Papers.* 1920-1926. 4 items.

T.L.S. (1926 June 2, Landing P. O., New Jersey), and 3 autograph signatures. MSS 980 A

1423. Röntgen, Wilhelm Conrad, 1845-1923. *Über die Elektriziatätsleitung in einigen Krystallen und über den Einfluss einer Bestrahlung darauf.* 1920. 218 [i.e., 224] p.: ill.; 36 cm.

T. ms. with holograph revisions, some several pages long, with colophon: München Frühjahr 1920; in German. Spine title: Roentgen, Manuscript, 1920. Includes folder of correspondence concerning the identification of the annotations as Röntgen's. MSS 1270 B

1424. Suess, Franz Eduard, 1867-1941. *Letter.* 1920. 1 item (4 p.)

A.L.S. (1920 Oct. 4, Vienna) to an unnamed correspondent on letterhead of the Geologisches Institut der Universität concerning the research of Prof. Kober; in German. MSS 1435 A

1425. Wirtinger, Wilhelm, 1865-1945. *Allgemeine Infinitesimalgeometrie in [--?--] Erfahrung von W. Wirtinger in Wien.* [192-?]. 1 item (30 leaves)

Holograph ms. (29, i.e., 30 leaves) with Latin quotation from Lucretius' *De rerum natura* pinned to leaf 12; in German. MSS 1577 A

1426. Bethenod, Joseph, 1883-1944. *Letter.* 1921. 1 item (2 p.): ill.

A.L.S. (1921 Feb. 20, Paris) to Roth concerning internal protection of turbo-generators; in French. MSS 109 A

1427. Brandes, Georg Morris Cohen, 1842-1927. *Letter.* 1921. 1 item (2 p.)

A.L.S. (1921 Sept. 7, Copenhagen) to Mrs. Eva Ingersoll Swazey quoting her father. MSS 168 A

1428. Coolidge, William David, 1873-1975. *Letter.* 1921. 1 item (1 p.)

A.L.S. (1921 March 13, Schenectady) on letterhead of the autograph collection of Howes Norris, Jr. MSS 368 A

1429. Hörbiger, Hanns, 1860-1931. *Papers.* 1921-1927. 6 items: ill.

Two A. transcriptions, one a poem, 1 T.L.S. with annotated T. abstract, 1 T.L., 2 architectural drawings for an observatory; in German. Additional material includes obituary notice (1931) and death announcement (1931).

MSS 707 A

1430. Huggins, Charles, 1901- *Letter.* [between 1921 and 1977]. 1 item (1 p.)

A.L.S., undated, thanking Mr. Silverman for a present.

MSS 730 A

1431. Prey, Adelbert, 1873-1949. *Letter.* 1921. 1 item (2 p.)

A.L.S. (1921 July 6, Prague) about astronomy and mentioning an event (1606 Oct. 31); in German.

MSS 1173 A

1432. Siedentopf, Henry Friedrich Wilhelm, 1872-1940. *Postcard.* 1921. 1 item (1 leaf)

A. postcard S. ([19]21 Aug. 24, [Berz-Dievenor?]) to Lakowitz; in German.   MSS 1366 A

1433. Brodetsky, Selig, 1888-1954. *Papers.* 1922-1934. 2 items: ill.

T.L.S. (1934 Dec. 11, Leeds) to Armin Weiner declining an invitation and enclosing a typed ms. article with handwritten annotations entitled: "The line of action of the resultant pressure in discontinuous fluid motion," published in *Proceedings of the AHW* (1922 Oct. 18).

MSS 182 A

1434. Eddington, Arthur Stanley, Sir, 1882-1944. *Papers.* 1922-1925. 2 items.

A.L.S. (1922 Feb. 19) concerning the effect of the speed of light upon digestion, and an autograph signature with epigram on letterhead of the Howes Norris, Jr., autograph collection, received 1925 May 11.   MSS 469 A

1435. Marconi, Guglielmo, marchese, 1874-1937. *Papers.* 1899-1934. 14 items.

Eight A.L.S. (one to Thompson), 2 T.L.S. (one in Italian), 1 A. postcard S., 1 A.S., and 2 telegrams, one from Thor Thörnblad; in English.   MSS 960 A

1436. Wigner, Eugene Paul, 1902- *Papers.* [between 1922 and 1977]. 7 items.

One T.L.S. (1964 Apr. 7, Princeton) to Karl Geyer, one holograph ms.: "The Unitary or Antiunitary Nature of Quantum Mechanical Variance Operations" (12 leaves), 1 holograph leaf: "The Nature of the Real," 1 typescript leaf S. (1971 Apr. 19): "Symmetries and Reflections," 1 holograph ms. (2 p.), and 2 holograph leaves, one numbered "17."   MSS 1569 A

1437. Bragg, William Lawrence, Sir, 1890-1971. *Papers.* 1923-1935. 4 items.

Three T.L.S. to Thirring concerning binding forces in crystal structures, and a holograph article entitled: "The New Crystallography." Additional materials include a copy of a letter from Thirring and a letter and envelope referring to the article.   MSS 162 A

1438. Gram, Christian, 1853-1938. *Letter.* 1923. 1 item (1 p.)

A.L.S. (1923 Feb. 21) to Norgaard; in Norwegian.

MSS 611 A

1439. Lenz, Oskar, 1848-1925. *Autograph signature.* 1923. 1 item (1 leaf)

Autograph signature (1923 April 19, [Baden-Looss?]) with Latin epigram.   MSS 882 A

1440. Millikan, Robert Andrews, 1868-1953. *Papers.* 1923-1936. 6 items.

Three T.L.S., one to Mrs. George M. Millard enclosing an A. ms. (4 p.) of a speech introducing Michelson, and one to Pupin, and 2 A.S.   MSS 1010 A

1441. Fog, Mogens Ludolf, 1904- *Autograph signature.* [between 1924 and 1977]. 1 item (1 p.)
MSS 523 A

1442. Warren, Charles, Sir, 1840-1927. *Letter.* 1924. 2 items.

A.L.S. ([19]24 May 15, Wiston) and A. envelope to Dr. Glaisher concerning Warren's book and mutual friends.

MSS 1533 A

1443. Borüvka, Otakar, 1899- *Papers.* [1925-1930?]. 3 items.

A.L.S. (1930 June 29, Paris); A. note with short biographies of some Czech mathematicians; and a holograph ms. entitled: "Sur les correspondances analytiques entre deux plans projectifs. II," published in a journal in 1927; in Czech and French.   MSS 146 A

1444. Byrd, Richard Evelyn, 1888-1957. *Letters.* 1925-1951. 4 items.

Four T.L.S. to various correspondents concerning Byrd's son and responding to letters of congratulation and encouragement.   MSS 207 A

1445. Daly, Reginald Aldworth, 1871-1957. *Letter.* 1925. 1 item (1 p.)

A.L.S. (1925 May 7, Harvard University) to Mr. Norris recommending the study of the earth on letterhead of the Howes Norris, Jr., autograph collection.   MSS 395 A

1446. Fabre, René, 1889-1966. *Letter.* 1925. 1 item (2 p.)

A.L.S. (1925 June 15) on letterhead of the Laboratoire de Toxicologie, Université de Paris concerning his proposed activity in industrial toxicology and hygiene; in French.   MSS 497 A

1447. Nernst, Walther, 1864-1941. *Papers.* 1925-1936. 5 items.

A.S. with Latin epigram. 2 A.L.S. and 1 envelope, and D.S.; in German.   MSS 1067 A

1448. Poulsen, Valdemar, 1869-1942. *Postcard.* 1925. 1 item (1 leaf)

A. postcard S. ([19]25 Apr. 19, Gjentofte, Denmark) to Werner Leschner; in German.   MSS 1166 A

1449. Snow, Charles Percy, 1905-1980. *Papers.* [between 1925 and 1977]. 3 items.

Three typescripts S.: 2 chapters (5, 9 p.) from *The Affair,* and 1 p. from *The Light and the Dark.*   MSS 1387 A

1450. Bethe, Han Albrecht, 1906- *Autograph signature.* [between 1926 and 1977]. 1 item (1 p.)

Card bearing autograph signature (n.d.).   MSS 108 A

1451. Born, Max, 1882-1970. *Papers.* 1926-1964. 3 items.

Two A.L.S. (1959 Sept. 20, and 1964 Jan. 17); and holograph ms. signed (1926 June 25) entitled: "Zur Quantenmechanik der Stossvorgaenge," later published in *Zeitscrift für Physik,* Berlin, 1926; in German and English.   MSS 144 A

1452. Chapman, Frank Michler, 1864-1945. *Letter.* 1926. 1 item (1 p.)

T.L.S. (1926 Sept. 7, New York) to Stitt concerning the question of introduced species on local bird lists.   MSS 330 A

1453. Cherrie, George Kruck, 1865-1946. *Letter.* 1926. 1 item (3 p.)

T.L.S. (1924 June 24, on board *S.S. Pan American* enroute for Rio de Janeiro) to Stitt on Stitt's A.L.S. (1926 May 31, Brookline, Mass.) requesting an autograph.   MSS 337 A

1454. Cushing, Harvey, 1869-1939. *Letters.* 1926. 2 items.

Two T.L.S. (1926 Feb. 3, and 1926 May 24, Boston) to Lane concerning a biography and a student rejected by Harvard Medical School.   MSS 388 A

1455. Ditmars, Raymond Lee, 1876-1942. *Letter.* 1926. 1 item (1 p.)

T.L.S. (1926 March 11, New York) on New York Zoological Park letterhead to Grinnel concerning The Living Natural History movie series.   MSS 444 A

1456. Funk, A. *Letter.* 1926. 1 item (2 p.)

A.L.S. (1926 Oct. 20, Warsaw) on letterhead of Panstswowa Szkoła Hygjeny asking about pictures of the trip from Stockholm to Uppsala and hoping to get to Prague in December; in French.   MSS 557 A

1457.　Kellogg, Vernon Lyman, 1867-1937. *Letter.* 1926. 1 item (1 p.)

T.L.S. (1926 Mar. 27, Washington, D.C.) to Rev. Frank Fitt concerning dogmatism in science.

MSS 779 A

1458.　Lindenthal, Gustav, 1850-1935. *Postcard.* 1926. 1 item (1 leaf)

A. postcard S. (1926 June 13, Vienna); in German.

MSS 899 A

1459.　Everling, Emil, 1890- *Letter.* 1927. 1 item (1 p.)

T.L.S. (1927 June 25, Berlin) to the *Berliner Tageblatt;* in German.　　　　　MSS 494 A

1460.　Ostwald, Wilhelm, 1853-1932. *Postcard.* 1927. 1 item (1 leaf)

A. postcard S. (1927 April 20, Grossbothen) to Herrn Mayer; in German.　　　MSS 1093 A

1461.　Sanger, Margaret 1879-1966. *Papers.* 1927-1929. 2 items.

T.L.S. (1929 May 29, New York City) to Mr. John F. Kendrick on letterhead of the Birth Control Clinical Research Center, and D.S. (1927 July 14) concerning the copyright to Sanger's *What Every Girl and Boy Should Know,* published by Brentano's. With 2 T.L. (carbon copies) signed by Arthur Brentano and Lowell Brentano concerning a related contract (1933 Feb. 3).　　MSS 1325 A

1462.　Weyl, Herman, 1885-1955. *Papers.* 1927-1928. 7 items.

Two A.L.S. and two A. envelopes to Armin Weiner, one holograph ms. (50 p.) plus one typescript with holograph equations and proofreading marks (46 leaves), both of "Quantenmechanik und Gruppentheorie," and one typescript of additions and corrections to that article (2 leaves); in German.　　　MSS 1555 A

1463.　Bush, Vannevar, 1890-1974. *Papers.* 1928-1964. 2 items: ill.

T.L.S. (1928 Feb. 24, Cambridge, Mass.) to Alfred C. Lane, concerning chabazite, and a signed first day cover (1964 Oct. 23) of the United Nations stamp commemorating cessation of nuclear testing.　　MSS 206 A

1464.　Cartan, Elie, 1869-1951. *Letter.* 1928. 1 item (4 p.)

A.L.S. (1928 July 17, Le Chesney) possibly to Borüvka concerning a paper Borüvka sent to Picard; in French.

MSS 307 A

1465.　Delépine, Stéphane Marcel, 1871-1965. *Letters.* 1928-1947. 16 items.

Fifteen A.L.S. and 1 T.L.S. to unidentified correspondents; in French.　　　MSS 423 A

1466.　Demolon, Albert, 1881-1954. *Letters.* 1928-1933. 4 items.

Four A.L.S. on letterhead of the Institut des recherches agronomiques; in French.　　MSS 426 A

1467.　Grignard, Victor, 1871-1935. *Papers.* 1928-1935. 8 items.

Five A.L.S., calling card S., holograph article "Sur la forme énolique de la Pulégone," published in the *Bulletin de la Société Chimique,* and 1 envelope; in French.

MSS 623 A

1468.　Heisenberg, Werner, 1901-1976. *Papers.* 1928-1936. 3 items.

Two A.L.S. to Thirring and Ehrenhaft, and a holograph ms. "Zur theorie des Ferromagnetismus," in German.　　　　MSS 680 A

1469.　Opel, Wilhelm. *An die Redaktion des Berliner Tageblatts, Berlin Betrifft: Veröffentlichungen "Mensch und Maschine."* 1928. 1 item (2 leaves)

A. ms. (1928 June 13, Rüsselsheim); in German.

MSS 1086 A

1470.　Roosevelt, Eleanor, 1884-1962. *Card.* [between 1928 and 1932]. 1 item (1 leaf)

A. card S. (Dec. 10, Executive Mansion, Albany, New York) to Mrs. O'Connor thanking her for a gift.

MSS 1230 A

1471.　Teller, Edward, 1908- *Autograph signature.* [between 1928 and 1977]. 1 item (1 leaf)

MSS 1451 A

1472. Arco, Georg, Graf von, 1869-1940. *Letter.* 1929. 1 item (1 p.)

T.L.S. (1929 Jan. 12, Berlin) from Arco to Rudolph Mosse-Haus. MSS 14 A

1473. Beck, Guido, 1903- *Letter.* 1929. 1 item (1 p.)

A.L.S. (1929 Dec. 7) concerning Heisenberg's work on ferromagnetism; in German. MSS 56 A

1474. Broglie, Louis, prince de, 1892- *Papers.* 1929-1945. 3 items.

Two A. cards S. (1929 Dec. 21 and 1945 May 18) and a holograph article entitled: "Sur une interprétation possible des conditions de stabilité de l'atome de Bohr," published in *Comtes-rendus de l'Académie des Sciences de Paris* in 1923; in French. Additional material includes newspaper clipping in German. MSS 185 A

1475. Dawson, Bertrand Edward Dawson, viscount, 1864-1945. *Letter.* 1929. 1 item (1 p.)

T.L.S. (1929 July 22) to Mr. Hodson concerning Lyttelton's writing on contraception and sexual perversion. Typed but unsigned response to Dawson's letter included in folder. MSS 412 A

1476. Franck, Heinrich, 1888- *Letter.* 1929. 1 item (1 p.)

T.L.S. (1929 Jan. 2, Berlin) to Zadek; in German. MSS 534 A

1477. Hantzsch, Arthur, 1857-1935. *Letters.* 1929-1932. 5 items.

Three A.L.S., and 2 A. postcards S. to Fajans; in German. MSS 662 A

1478. Keith, Arthur, Sir, 1866-1955. *Letter.* 1929. 1 item (1 p.)

A.L.S. (1929 July 18, London) to Hodson concerning birth control. MSS 776 A

1479. Schrödinger, Erwin, 1887-1961. *Papers.* 1929-1936. 6 items.

Two A.L.S. (1936), 1 A.L.S. (1934 Nov. 16) to Hans Thirring, 1 A. envelope (1929) to Armin Weiner, all in German, 1 A.L.S. (1934 May 18) to Howes Norris on letterhead of "The Autograph Collection of Howes Norris, Jr.," in English, and 1 autograph ms. (folios 12 and 12a only) citing the work of Albert Einstein; in German. MSS 1337 A

1480. Wegscheider, Rudolph, 1859-1935. *Card.* 1929. 1 item (1 leaf)

Printed card (1929 Oct., Vienna) with holograph greeting, closing, and signature, to a "Frau Doktor"; in German. MSS 1541 A

1481. Zeeman, Pieter, 1865-1943. *Letters.* 1929-1930. 3 items.

Two A.L.S. (1929 Nov. 23 and 1930 Nov. 1, Amsterdam) to Ehrenhaft, and A. card S. (1930 Dec. 17, Amsterdam) to Eile(?); in German. MSS 1594 A

1482. Beebe, Charles William, 1877-1962. *Letter.* 1930. 1 item (1 p.)

A.L.S. on letterhead of Bermuda Oceanographic Expedition requesting material from a librarian. MSS 83 A

1483. Bell, Eric Temple, 1883-1960. *Formula.* 1930. 1 item (1 p.)

A. formula S. (1930 Oct. 27). MSS 87 A

1484. Mather, Kirtley Fletcher, 1888-1978. *Letter.* 1930. 1 item (1 leaf)

T.L.S. (1930 Dec. 19, Cambridge, Mass.) to Grace L. West concerning a Sunday evening forum. MSS 974 A

1485. Monod, Jacques Lucien, 1910-1976. *Fragment.* [between 1930 and 1976]. 1 item (1 leaf)

A. ms. fragment titled "Results." MSS 1026 A

1486. Thum, August, 1881-1957. *Letter.* 1930. 1 item (1 leaf)

T.L.S. (1930 Nov. 7, Darmstadt) to his editor, Walter Zadek, about Thum's "Nutzen der technischwissenschaftlichen Forschung"; in German. MSS 1467 A

1487. Boutaric, Augustin Marius Arsène, 1885- *Letters.* 1931-1932. 2 items.

Two A.L.S. (1931 Dec. 12, and 1932 April 24, Dijon) recommending young scientists to his correspondents; in French.   MSS 158 A

1488. Courant, Richard, 1888-1972. *Letter.* 1931. 2 items.

T.L.S. (1931 March 1, Göttingen) with envelope; to A. Weiner; in German.   MSS 377 A

1489. Crozier, William, 1855-1942. *Letter.* 1931. 1 item (1 p.)

T.L.S. (1931 Mar. 12, Washington, D.C.) to Strasser concerning an important decision in Crozier's life.   MSS 386 A

1490. Dingeldey, Friedrich, 1859-1939. *Document.* 1931. 1 item (1 p.)

A.D.S. (1931 July 15, Darmstadt) confirming that a student attended lectures of higher mathematics; in German.   MSS 443 A

1491. Duryea, Charles E., 1861-1938. *Letter.* 1931. 1 item (1 p.)

T.L.S. (1931 Jan. 31, Philadelphia) to Mrs. Thwing concerning a pamphlet on Morey. Additional material on Duryea cars.   MSS 465 A

1492. Karpinski, Louis Charles, 1878-1956. *Letters.* 1931-1937. 10 items.

Eight A.L.S., 1 A. note S., and 1 addressed envelope, all to Gold.   MSS 769 A

1493. Linke, Karl Wilhelm Franz, 1878-1944. *Letters.* 1931-1934. 4 items.

Four A.L.S.; in German.   MSS 902 A

1494. Muller, Hermann Joseph, 1890-1967. *Papers.* 1931-1948. 3 items.

T.L.S. (1948 Oct. 9, Bloomington, Ind.), A.D.S. (1931 May 18, Austin, Texas) on letterhead of the Autograph Collection of Howes Norris, Jr., and A.S. with short typed biography.   MSS 1047 A

1495. Pearson, Karl, 1857-1936. *Letter.* 1931. 1 item (1 leaf)

A.L.S. (1931 July 26) concerning eugenics on letterhead of the Francis Galton Eugenics Laboratory, University of London.   MSS 1106 A

1496. Piccard, Auguste, 1884-1962. *Letters.* 1931-1936. 3 items.

A.L.S. (1931 Aug. 9, Praz-de-Fort, Wallis, i.e., Valais) to Julius Renk, with envelope, and T.L.S. (1936 Apr. 13) to a London news agency, both about Piccard's balloon ascents; in German. Additional material includes Renk's copy of a newspaper article about Piccard.   MSS 1139 A

1497. Richardson, Owen Willans, 1879-1959. *Papers.* 1931. 5 items.

A.L.S. (1931 April 7, Haverstock Hill, Hampstead) to an unnamed colleague in Vienna, in English; 2 T.L.S. (1931 Oct. and Nov. 3, Vienna) from Ehrenhaft to A. Weiner, all three letters concerning Richardson's lecture in Vienna; 1 ticket for that lecture (1931 Oct. 27, Vienna), and 1 copy of an article about Richardson with a photographic portrait; in German.   MSS 1210 A

1498. Wilkins, George Hubert, Sir, 1888-1958. *Letter.* 1931. 1 item (1 leaf)

T.L.S. (1931 July 8, Devonport) to Gerald Christy on letterhead of the "Trans-Arctic Submarine Expedition, Inc., conducting the Wilkins-Ellsworth Trans-Arctic Submarine Expedition," concerning a lecture tour in London.   MSS 1571 A

1499. Borsig, Ernst von, 1906- *Letter.* 1932. 1 item (1 p.)

T.L.S. (1932 May 3, Gross-Behnitz) to Zentralredaktion für Deutsche Zeitungen; in German.   MSS 145 A

1500. Chapman, Frederick Spencer, 1907-1971. *Letter.* [1932]. 1 item (2 p.)

A.L.S. ([1932] Jan. 28, London) to Mr. Christy declining most lecture offers because he is working hard on the book about the Greenland expedition.   MSS 331 A

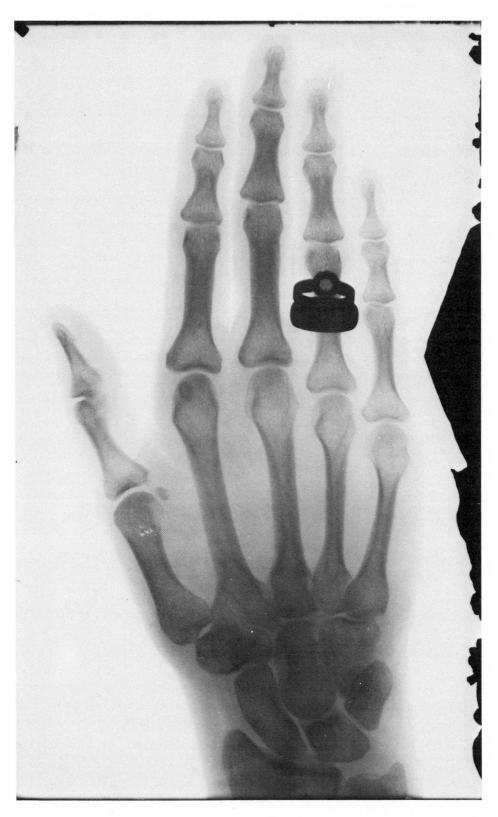

XV.   An early X-ray made by Wilhelm Röntgen of a child's hand
      (1897).
      Entry No. 1128, SI Neg. 84-10402.

1501.   Duisberg, Carl, 1861-1935. *Letter.* 1932. 1 item (1 p.)

T.L.S. (1932 Mar. 10) to the Zentralredaktion für Deutsche Zeitungen; in German.   MSS 461 A

1502.   Ehrenhaft, Felix, 1879-1952. *Letters.* 1932-1947. 2 items.

A.L.S. (1932 March 22, Vienna), and A. card S. (1947 April 16, Vienna); in German.   MSS 474 A

1503.   Robinson, Wirt, 1864-1929. *Letters to Robinson.* 1894-1941. 20 items.

Sixteen A.L.S. (1894-1902), three with A. envelopes, to Robinson from Allen, Bangs, Bendire, Brewster, Garman, Henshaw, Lucas, Miller, Scudder, Stone, Strecker, Taylor, and Thaxter, and T.L.S. (1941 Nov. 6) to Mrs. Robinson from Wetmore.   MSS 1226 A

1504.   Gerlach, Walther, 1889- *Papers.* 1932-1961. 7 items.

Three T.L.S. and an A.L.S. to Laub, typed article with handwritten corrections titled "Versuch eines möglichst einheitlichen physikalischen Weltbildes. . .", and 2 inscribed offprints; in German.   MSS 587 A

1505.   Honnef, Hermann. *Letter.* 1932. 1 item (1 p.)

T.L.S. (1932 Feb. 23, Berlin) to Ernst Erwin about Honnef's high-power windmill; in German.   MSS 714 A

1506.   Perrot, Emile, 1867- *Letter.* 1932. 1 item (2 p.)

A.L.S. (1932 Apr. 10, Paris) on letterhead of the Faculté de Pharmacie, Université de Paris; in French.   MSS 1120 A

1507.   Stefansson, Vilhjalmur, 1879-1962. *Letters.* 1932-1948. 5 items.

A. postcard S. and 4 T.L.S. to Helen M. Gunz (concerning cheyne), Professor Mason, and Bernice Reich.   MSS 1410 A

1508.   Carnot, Paul, b. 1869. *Letter.* 1933. 1 item (1 p.)

A.L.S. (1933 Nov. 10, Paris) recommending Dr. Ucko; in French.   MSS 303 A

1509.   Fricke, Hermann, 1876-1949. *Letter.* 1933. 1 item (1 p.)

A.L.S. (1933 Mar. 17, Berlin) to the editors of *Deutsche Zeitung;* in German.   MSS 548 A

1510.   Hess, Victor Francis, 1883-1964. *Papers.* 1933-1961. 5 items.

Two A.L.S., 1 T.L.S., holograph ms. of notes (3 p.), and holograph ms. of calculations (2 p.); in German and English.   MSS 698 A

1511.   Lévi, Sylvain, 1863-1935. *Letter.* 1933. 1 item (1 leaf)

A.L.S. (1933 Oct. 18, Paris) to an unnamed correspondent concerning a position in industry for Pringsheim; in French.   MSS 888 A

1512.   Poulenc, Camille, b. 1864. *Letter.* 1933. 1 item (3 p.)

A.L.S. ([19]33 May 30, Paris) recommending a young man; in French.   MSS 1164 A

1513.   Tiffeneau, Marc Emile Pierre Adolphe, 1873-1945. *Letter.* 1933. 1 item (2 leaves)

A.L.S. (1933 June 2, Paris) to a friend on letterhead of the Laboratoire de Pharmacologie et de Matière Médicale, Université de Paris, introducing Kraus; in French.   MSS 1469 A

1514.   Aston, Francis William, 1877-1945. *Papers.* 1934. 4 items.

A.L.S. (1934 July 1, Cambridge, England) from Aston to Dr. Weiner (?) enclosing a rough pencil draft of a letter (1934 Feb. 17) from Aston to *Nature* on the composition of rare earth metals, and 2 envelopes addressed to Dr. Weiner, (1934 June 27, 1934 July 1). Typed transcription of letter to *Nature* and 3 carbons in folder.   MSS 29 A

1515.   Darier, Jean, 1856-1938. *Letter.* 1934. 1 item (3 p.)

A.L.S. (1934 Jan. 25, Paris) to an unnamed correspondent recommending Rudolph Stern for a list of German refugee scientists and doctors, on the letterhead of the Académie de Médicine; in French.   MSS 402 A

1516. Lespieau, Robert, 1864-1947. *Letter.* 1934. 1 item (4 p.)

A.L.S. (1934 Dec. 16, Paris) to an unnamed correspondent concerning the work of Gredy and Kohlrausch.
       MSS 886 A

1517. Charcot, Jean Baptiste Auguste Etienne, 1867-1936. *Letter.* 1935. 1 item (2 p.)

A.L.S. (1935 March 2, Neuilly-sur-Seine) to Du Puigaudeau concerning her situation with the Société de Géographie, with envelope; in French.   MSS 334 A

1518. Compton, Arthur Holly, 1892-1962. *Papers.* 1935-1959. 6 items.

Three T.L.S., autograph card, holograph article S. (12 p.) entitled "Cosmic Rays," published in *Nature,* May 4, 1935, and envelope. Photocopy and offprint of article included.   MSS 364 A

1519. Hoover, Herbert, 1874-1964. *Papers.* 1935-1938. 4 items.

Three T.L.S., one incomplete, and a printed pamphlet with envelope.   MSS 719 A

1520. Leclainche, Emmanuel, 1861-1953. *Letter.* 1935. 1 item.

A.L.S. (1935 April 29) to an unidentified correspondent recommending a student, Mlle. Brigando; in French.
       MSS 868 A

1521. Perrin, Jean Baptiste, 1870-1942. *Papers.* 1935-1936. 4 items.

T.L. (carbon copy) S., mimeographed L.S., and 2 passes signed by Perrin, all concerning the Palais de la Découverte of the Exposition Internationale de Paris, 1937; in French.   MSS 1127 A

1522. Siegbahn, Manne, 1886-1978. *Letter.* 1935. 2 items.

A.L.S. (no date) and A. envelope ([19]35 May 9 or Sept. 5) to Armin Weiner. The letter, written on the verso of a leaf of ms., identifies the text on the recto as "Zur Spektroskopie der ultraweichen Röntgenstrahlung III"; in German.   MSS 1367 A

1523. Dafoe, Allan Roy, 1883-1943. *Letter.* 1936. 1 item (2 leaves); ports.

A.L.S. (1936 Feb. 10, Callender, Ontario) to Jose D. Carpio concerning the Dionne Quintuplets on Dionne Quintuplet Guardianship letterhead.   MSS 391 A

1524. Stevens, John Frank, 1853-1943. *Letter.* 1936. 1 item (2 p.)

A.L.S. ([19]36 Mar. 27, Canal Zone) to Prof. Carpio on letterhead of the Hotel Tivol, Ancon, C.Z., citing his series of articles, "An Engineer's Recollections."   MSS 1417 A

1525. Urbain, Georges, 1872-1938. *Letters.* 1936-1938. 16 items.

Sixteen T.L.S. (some mimeographed, 1 carbon) from Urbain as Président de la Section de Chimie, Palais de la Découverte, to colleagues, some annotated; in French.
       MSS 1490 A

1526. Von Mises, Richard, 1883-1953. *Papers.* 1936. 5 items.

A.L.S. (1936 Aug. 10, Vienna) with envelope, T.L.S. (1936 Sept. 12, Istanbul) with envelope, A. ms. S. (8 leaves) "Segelflug und Aehnlichkeitsgesetz;" in German. Offprint of article published in *Zeitschrift für angewandte Mathematik und Mechanik.*   MSS 1014 A

1527. Douglas, A. E. *Letter.* 1937. 1 item (1 p.)

T.L.S. (1937 Oct. 11, Tucson) to Baumgart concerning eclipse pictures.   MSS 450 A

1528. Fallot, Paul, 1889-1960. *Letter.* 1937. 1 item (1 p.)

A.L.S. (1937 Dec. 3) on letterhead of the Institut de Géologie Appliquée, Nancy, concerning candidates for the chair of geology at the College of France; in French.
       MSS 500 A

1529. Gay, Louis Maxime Léon, 1881-1958. *Letter.* 1937. 1 item (2 p.)

A.L.S. (1937 June 8, Montpellier) concerning a speaking invitation and the return of his mentor, Prof. Godechot (?), on letterhead of the Faculté des Sciences de Montpellier, Institut de Chimie; in French.   MSS 577 A

1530. Jolibois, Pierre, 1884-1954. *Letter.* 1937. 1 item (1 p.)

T.L.S. (1937 June 8, Paris) to the President of the Commission des bourses de Recherches de Chimie, Collège de France, concerning the appointment of Bossuet as technical aide in Jolibois' laboratory; in French. MSS 757 A

1531. Pascal, Paul Victor Henri, 1880-1968. *Letter.* 1937. 1 item (2 p.)

A.L.S. (1937 June 8, Paris) on letterhead of the Laboratoire de chimie minérale, Faculté des sciences de Paris; in French. MSS 1098 A

1532. Rouch, Jules Alfred Pierre, 1884- *Letter.* 1937. 1 item (1 leaf)

A.L.S. (1937 April 17, Toulon) to an unnamed professor seeking his support for Rouch's candidacy for the vacant chair of oceanographic physics at the Institut Océanographique; in French. MSS 1237 A

1533. Charpentier, Dr. *Letter.* 1938. 1 item (1 p.)

A.L.S. (1938 Feb. 28, Dijon) concerning funding of research on proteins and antibodies in serum; in French. MSS 335 A

1534. Garretson, Martin S. *Letter.* 1938. 1 item (1 p.)

A.L.S. (1938 June 13) to Zabriskie concerning copies of *The History of the Bison* to be bound in buffalo leather. MSS 569 A

1535. Waksman, Selman Abraham, 1888-1973. *Letters.* [between 1938 and 1973]. 2 items.

T.L.S. (1945 Dec. 17, New Brunswick, N. J.) to Dr. Schymen Nussbaum on letterhead of the Agricultural Experiment Station, and autograph signature cut from a T.L.S. MSS 1521 A

1536. De Forest, Lee, 1873-1961. *Letters.* 1939. 2 items.

Two A.L.S. (1939 Feb. 21, 1939 May 15, Los Angeles) to Goldman with thanks for his concerts and spoken tribute to De Forest, and regrets at not being able to meet. MSS 417 A

1537. Jacobson, Herbert Lawrence, 1915- *Autograph signature.* [between 1940 and 1973]. 1 item (1 p.). MSS 745 A

1538. Jennings, Allyn R. *Letters.* 1940-1941. 2 items.

Two T.L.S. (1940 Aug. 13, and 1941 Dec. 17) to Zabriskie on letterhead of the Director, New York Zoological Park. MSS 754 A

1539. Delbet, Pierre Louis Ernest, 1861-1957. *Letter.* 1942. 1 item (1 p.)

A.L.S. (1942 Mar. 4) to an unnmamed correspondent concerning molecules and metalloids; in French. MSS 422 A

1540. Stanley, Wendell Meredith, 1904-1971. *Papers.* [between 1942 and 1971]. 2 items.

A.L.S. (1942 Dec. 10) to Norris on letterhead of The Autograph Collection of Howes Norris, Jr., and autograph signature on brief T. summary of his research leading to the Nobel Prize in 1946. MSS 1405 A

1541. Čech, Eduard, 1893-1960. *Ricerche sulla teoria delle congruenze e dei complessi di rette memoria di Guido Fubini.* [between 1943 and 1960]. 1 item (18 leaves)

Holograph ms. signed. MSS 318 A

1542. Stern, Otto, 1888-1969. *Autograph signature.* [between 1943 and 1969]. 1 item (1 leaf)

Autograph signature on brief T. summary of his research leading to the Nobel Prize with Stern's holograph revisions of the T. text. MSS 1416 A

1543. Weiner, Armin, b. 1880. *Papers.* 1943-1963. 10 items: port.

Five T.L.S. (1943-1952) to Dibner, two A.L.S. (1960 and 1963) from Weiner's daughter, Joan Hoffman, to Dibner, two copies (original and carbon) of Weiner's curriculum vitae, and 1 autograph signature on a photographic portrait. MSS 1543 A

1544. Rabi, Isidor Isaac, 1898- *Papers.* [between 1944 and 1977]. 2 items.

T. sheet S. (1969) with text from an article in the *New York Post* on the moon landing and with the "First Man on the Moon" stamp with first day of issue cancellation and cancellation dates July 20 and Sept. 9, and autograph signature on brief T. description of his work leading to the Nobel Prize in 1944.   MSS 1184 A

1545. Fleming, Alexander, Sir, 1881-1955. *Letter.* 1945. 1 item (1 p.)

A.L.S. (1945 July 5) concerning Norris' autograph collection on letterhead of the Howes Norris, Jr., autograph collection.   MSS 520 A

1546. Whittaker, Edmund Taylor, 1873-1956. *Letter.* 1945. 1 item (1 leaf)

A.L.S. (1945 Dec. 13, 48 George Square, Edinburgh) to Richardson concerning a letter and a ms. and expressing sympathy for the death of his wife.   MSS 1561 A

1547. Briner, Hermann, 1918- *Amplituden- und frequenz-modulation.* 1947. 1 item (12 p.): ill.

T. article (1947 Oct.) with handwritten formulae; in German. Sheet of diagrams (Fig. 3) laid in.   MSS 179 A

1548. Chevenard, Pierre, 1888-1960. *Letter.* 1947. 1 item (1 p.)

T.L.S. (1947 Feb. 25, Imphy) declining an invitation to the first reunion of the Comité Langevin and suggesting names to be added to a list of scientist victims of the Germans; in French.   MSS 338 A

1549. Magrou, Joseph, 1883-1951. *Letter.* 1947. 1 item (1 leaf)

A.L.S. (1947 Jan. 7, Paris) to an unnamed correspondent concerning M. Dauvillier, on letterhead of the Institut Pasteur; in French.   MSS 950 A

1550. Murphy, Robert Cushman, 1887-1973. *Letter.* 1947. 1 item (1 leaf)

T.L.S. (1947 July 30) with ms. postscript to Zabriskie concerning an exhibition on letterhead of The American Museum of Natural History.   MSS 1051 A

1551. Russell, Edward John, Sir, 1872-1965. *Letter.* 1947. 1 item (2 p.)

A.L.S. (1947 Mar. 6, Campsfield Wood, Woodstock, Oxon.) to "mon cher maitre" thanking him for his assisting in Russell's election as "Associé Etranger de l'Académie des Sciences"; in French.   MSS 1245 A

1552. Einstein, Albert, 1879-1955. *Relativity.* 1948. 1 item (9 leaves)

Two leaves of typescript and 7 leaves of galley proofs with corrections in Einstein's hand for the article "Relativity" in *American Peoples Encyclopedia.*   MSS 248 A

1553. Bohr, Niels Henrik David, 1885-1962. *Document.* 1949. 1 item (1 p.)

D.S. (1949 Jan. 5) concerning Augusta Rasmussen; in Danish.   MSS 134 A

1554. Cooper, Leon N., 1930- *Autograph card.* [between 1950 and 1977]. 1 item (1 p.)   MSS 370 A

1555. Kendall, Edward Calvin, 1886-1972. *Document.* [between 1950 and 1972]. 1 item (2 p.)

T. biographical sketch S. (undated).   MSS 781 A

1556. Moulton, Forest Ray, 1872-1952. *Letter.* 1951. 1 item (1 leaf)

A.L.S. (1951 Aug. 31) to Bettie Smith concerning photographs he is sending.   MSS 1044 A

1557. Perrin, Francis, 1901- *Letter.* 1951. 1 item (1 leaf)

T.L.S. (1951 Oct. 31, Paris) to Gignoux seeking admission to the Académie des Sciences; on letterhead of the Laboratoire de Physique Atomique et Moléculaire, Collège de France; in French.   MSS 1117 A

1558. Eisenhower, John Sheldon Doud, 1922- *Letter and photograph.* 1952. 4 items; ports.

A.L.S. (1952 Oct. 4, Korea) to Frank Beebe, and 3 signed photographs of Dwight D. Eisenhower, Mamie Eisenhower, and John Eisenhower.   MSS 477 A

1559. Olivecrona, Herbert, 1891- *Letter.* 1952. 1 item (1 leaf)

T.L.S. (1952 Sept. 24, Stockholm) to Herr Knud Capozzi-Bentzen; in Swedish. MSS 1085 A

1560. Purcell, Edward M., 1912- *Autograph signature.* [between 1952 and 1977]. 1 item (1 leaf)

Autograph signature on page with short typed summary of his work leading to the Nobel Prize in 1952. MSS 1181 A

1561. Vandenberg, Arthur Hendrick, 1907-1968. *Letter.* 1952. 1 item (1 leaf)

T.L.S. (1952 July 28, Denver) to Frank J. Beebe on letterhead of the Office of Dwight D. Eisenhower acknowledging, on behalf of General Eisenhower, Beebe's letter. MSS 1498 A

1562. Dessauer, Friedrich, b. 1881. *Papers.* 1953-1959. 29 items.

Six A. postcards S., 1 A.L.S., 17 T.L.S., chiefly addressed to Laub, 1 signed calling card, and 2 corrected carbon copies each of 2 articles; in German. Additional picture postcard signed Dr. O. Zuppinger. MSS 435 A

1563. Lipmann, Fritz Albert, 1899- *Papers.* 1953-1970. 6 items.

Three A.S. attached to typed sheets, 1 page of typed ms. with holograph corrections and additions, 1 A. note S., all in English, and 1 A. postcard S. to Loewi, in German. MSS 905 A

1564. Chadwick, James, 1891-1974. *Letters.* 1954-1963. 3 items.

Two A.L.S., one concerning the dangers of atomic radiation, and 1 T.L.S. MSS 322 A

1565. Seaborg, Glenn Theodore, 1912- *Papers.* 1954-1965. 6 items.

Four T.L.S. (1954-1961) to Paul Shank, Bettie Smith, and Hans Mueller, and 2 autograph signatures, one on a brief T. summary of his research leading to the Nobel Prize in 1951. MSS 1345 A

1566. Weller, Thomas Huckle, 1915- *Autograph signatures.* [between 1954 and 1977]. 2 items.

Autograph signatures on calling card and on brief T. summary of his research leading to the Nobel Prize in 1954. MSS 1550 A

1567. Domagk, Gerhard, 1895-1964. *Papers.* 1955. 2 items: port.

T.L.S. (1955 Aug. 31, Wuppertal-Elberfeld), and inscribed photograph; in German. MSS 448 A

1568. Robbins, Frederick C., 1916- *Autograph signature.* [between 1955 and 1977]. 1 item (1 leaf)

Autograph signature (undated) on page with brief T. description of his work leading to the Nobel Prize in 1954. MSS 1220 A

1569. Salk, Jonas Edward, 1914- *Scientific articles.* [between 1955 and 1977]. 2 items.

T. ms. S. (4 p.): "Some Characteristics of a Continuously Propagating Cell Derived from Monkey Heart Tissue," and T. ms. (carbon copy) S. (7 p.): "Vaccines for Poliomyelitis, [copied from] *Scientific American*, April 1955." Salk added the words "copied from" to the 2nd title. MSS 1321 A

1570. Untermeyer, Louis, 1885-1977. *Albert Einstein.* [between 1955 and 1977]. 1 item (1 leaf)

Signed typescript of a chapter from Untermeyer's *Makers of the Modern World.* MSS 1488 A

1571. Urey, Harold Clayton, 1893-1981. *Papers.* 1955-1967. 6 items.

Two T.L.S. (1955, 1966), 1 autograph signature on 3 x 5 card, 1 autograph signature on a First Day Cover (1967 Aug. 1), 1 autograph signature on a brief T. summary of his research leading to the Nobel Prize in 1934, and possibly autograph signature (surname only, on envelope, 1965). MSS 1492 A

1572. Wallace, Henry Agard, 1888-1965. *Letters.* 1956-1958. 4 items.

Four T.L.S. to Dibner on letterhead of Farvue Farm, South Salem, New York, concerning corn experiments, DNA, and other scientific topics. MSS 1527 A

1573.   Guinier, Philibert, 1876-1962. *Letter.* 1957. 1 item (2 p.)

A.L.S. (1957 May 27, Paris) to an unidentified correspondent with congratulations and excuses for not attending the reception; in French.   MSS 634 A

1574.   Pérès, Joseph, 1890-1962. *Card.* 1957. 1 item (2 p.)

A. card S. ([19]57 May 27, Paris) to a colleague on his ninetieth birthday; on letterhead of Le Doyen de la Faculté des Sciences, Université de Paris; in French.   MSS 1114 A

1575.   Sikorsky, Igor Ivanovich, 1889-1972. *Papers.* 1958-1971. 4 items.

One T.L.S. (1958 Feb. 3, Bridgeport, Connecticut) to Mr. Charles A. Pearce, and 3 T. mss. S.: 2 drafts of "The First Helicopter" and 1 of "The Helicopter Enters Regular Service."   MSS 1372 A

1576.   Tatum, Edward Lawrie, 1909-1975. *Autograph signature.* [between 1958 and 1977]. 1 item (1 leaf)

Autograph signature on brief T. summary of his research leading to the Nobel Prize in 1958.   MSS 1445 A

1577.   Groves, Leslie R., 1896-1970. *Letter and photograph.* 1959. 2 items: port.

T.L.S. (1959 Feb. 13, Stamford, Conn.) to Paul Shank concerning Groves' views on atomic energy, and an inscribed photograph.   MSS 627 A

1578.   Windaus, Adolf, 1876-1959. *Letter.* 1959. 1 item (1 leaf)

A.L.S. ([19]59 Feb. 5, Göttingen) to Herr Mertens; in German.   MSS 1575 A

1579.   Bloch, Felix, 1905- *Autograph signature.* [between 1960 and 1977]. 2 items.

Autograph signature and envelope addressed to Gary A. Woolson.   MSS 121 A

1580.   Chandrasekhar, Subrahmanyan, 1910- *Autograph card.* 1960. 1 item (1 p.)

Autograph signature with mathematical formula on card (1960 Oct. 11, Williams Bay, Wis.).   MSS 328 A

1581.   Hillary, Edmund, Sir, 1919- *Summit.* [between 1960 and 1973]. 1 item (12 leaves)

T. ms. S. concerning the conquest of Everest.   MSS 703 A

1582.   Pauling, Linus, 1901- *Papers.* 1960-1975. 9 items.

A.L.S.; 2 A.S., one on a first day cover commemorating Albert Einstein; 4 T.L.S.; and 2 T. ms. S., one with a first day cover commemorating the U. N. Nonproliferation of Nuclear Weapons Agreement.   MSS 1104 A

1583.   Brattain, Walter H., 1902- *Papers.* 1962. 4 items.

Two typed biographical extracts signed, 1 typed extract from Nobel lecture of 1956 signed, and A.L.S. (1962 Nov. 13) advising Mr. Eisner to spell names correctly.   MSS 174 A

1584.   Glenn, John, 1921- *Three to make ready: a detailed plan.* [between 1962 and 1977]. 1 item (17 leaves)

T. article S. concerning his experiences in preparing for his orbital flight.   MSS 599 A

1585.   Watson, James D., 1928- *Autograph signature.* [between 1962 and 1977]. 1 item (1 leaf)
                                    MSS 1535 A

1586.   Alvarez, Luis Walter, 1911- *Autograph signatures.* 1963. 2 items.

One signature on first day cover, National Academy of Sciences (1963 Oct. 14); one on plain slip, undated.   MSS 75 A

1587.   Chamberlain, Owen, 1920- *Papers.* 1963. 2 items.

T.L.S. (1963 Mar. 22, Berkeley) regretting that he has no photographs to send, and an autograph signature.   MSS 325 A

aus deren Verjüngung $^{*}G_{\mu\nu}$ (92,42), beziehungsweise aus dem symmetrischen und dem antisymmetrischen Bestandteil dieses Tensors:

$$\gamma_{\mu\nu} = -\frac{\partial \Gamma^{\alpha}_{\mu\nu}}{\partial x_{\alpha}} + \Gamma^{\alpha}_{\mu\beta}\Gamma^{\beta}_{\nu\alpha} + \frac{1}{2}\left(\frac{\partial \Gamma^{\alpha}_{\mu\alpha}}{\partial x_{\nu}} + \frac{\partial \Gamma^{\alpha}_{\nu\alpha}}{\partial x_{\mu}}\right) - \Gamma^{\alpha}_{\mu\nu}\Gamma^{\beta}_{\alpha\beta} \cdots (2)$$

$$\varphi_{\mu\nu} = \frac{1}{2}\left(\frac{\partial \Gamma^{\alpha}_{\mu\alpha}}{\partial x_{\nu}} - \frac{\partial \Gamma^{\alpha}_{\nu\alpha}}{\partial x_{\mu}}\right). \cdots (3)$$

Gemäss dieser Voraussetzung erhält man an Stelle von (1) zunächst

$$\int (y^{\mu\nu}\delta\gamma_{\mu\nu} + f^{\mu\nu}\delta\varphi_{\mu\nu})d\tau = 0, \cdots (1a)$$

wobei gesetzt ist

$$\left.\begin{array}{l}\frac{\partial \mathfrak{L}}{\partial \gamma_{\mu\nu}} = y^{\mu\nu} \\[2mm] \frac{\partial \mathfrak{L}}{\partial \varphi_{\mu\nu}} = f^{\mu\nu}\end{array}\right\} \cdots (4)$$

In (1a) sind $\delta\gamma_{\mu\nu}$ und $\delta\varphi_{\mu\nu}$ vermöge (2) und (3) durch die $\Gamma^{\sigma}_{\mu\nu}$ und $\delta\Gamma^{\sigma}_{\mu\nu}$ aus-zudrücken. Mit Rücksicht darauf, dass die 40 Variationen $\delta\Gamma^{\sigma}_{\mu\nu}$ von einander unabhängig wählbar sind, erhält man aus (1a) vierzig Gleichungen:

$$y^{\mu\nu}_{;\alpha} - \frac{1}{2}y^{\mu\sigma}_{;\sigma}\delta^{\nu}_{\alpha} - \frac{1}{2}y^{\nu\sigma}_{;\sigma}\delta^{\mu}_{\alpha} - \frac{1}{2}i^{\mu}\delta^{\nu}_{\alpha} - \frac{1}{2}i^{\nu}\delta^{\mu}_{\alpha} = 0. \cdots (1b)$$

Dabei sind die Tensordichten

$$y^{\mu\nu}_{;\alpha} = \frac{\partial y^{\mu\nu}}{\partial x_{\alpha}} + y^{\sigma\nu}\Gamma^{\mu}_{\sigma\alpha} + y^{\mu\sigma}\Gamma^{\nu}_{\sigma\alpha} - y^{\mu\nu}\Gamma^{\sigma}_{\alpha\sigma} \cdots (5)$$

$$i^{\mu} = \frac{\partial f^{\mu\sigma}}{\partial x_{\sigma}} \cdots (6)$$

eingeführt. Die 40 Gleichungen (1b) erlauben uns, die 40 Grössen $\Gamma^{\sigma}_{\mu\nu}$ durch die $y^{\mu\nu}$, $f^{\mu\nu}$ und deren Ableitungen auszudrücken. Um dies zu bewerk-stelligen, muss man von den kontravarianten Tensordichten zu den kontravarianten Tensoren und von diesen zu den kovarianten Tensoren übergehen. Wir definieren zu diesem Zweck die Tensoren $g^{\mu\nu}$ und $g_{\mu\nu}$ durch die Gleichungen

$$\left.\begin{array}{l}g^{\mu\nu}\sqrt{-g} = y^{\mu\nu} \\[1mm] g_{\mu\sigma}g^{\nu\sigma} = \delta^{\nu}_{\mu} \\[1mm] g = |g_{\mu\nu}|,\end{array}\right\} \cdots (7)$$

XVI.   A page of notes and calculations of Albert Einstein (ca. 1930).
Entry No. 1353, SI 82-4785.

1588. Shockley, William, 1910- *Autograph signatures.* 1963-1973. 3 items.

Two autograph signatures, one with sketch, on first day covers (1963 Oct. 14, 100th Anniversary of the National Academy of Science, and 1973 July 10, Progress in Electronics), and autograph signature ([19]66 Aug. 21) on brief T. summary of his research leading to the Nobel Prize in 1956. MSS 1365 A

1589. Wald, George, 1906- *Autograph signature.* 1963. 1 item (1 leaf)

Autograph signature and greeting on First Day Cover (1963 Oct. 14, Washington, D.C.) for the National Academy of Sciences of the United States of America. MSS 1523 A

1590. Nier, Alfred Otto Carl, 1911- *Letter.* 1964. 1 item (1 leaf)

T.L.S. (1964 Jan. 17, Minneapolis) to Patricia Burkett concerning her school project. MSS 1074 A

1591. Funk, Casimir, 1884-1967. *Letter.* 1965. 2 items.

T.L.S. (1965 Aug. 10, New York) to Americans Seeking Knowledge promising information on his life work, with envelope. MSS 558 A

1592. Grissom, Virgil I., 1926-1967. *Letters.* 1965-1966. 2 items.

T.L.S. (1965 April 29) to George McCullough concerning a flight in "Molly Brown," and a T.L.S. (1966 Jan. 21) to Bern Dibner concerning the authenticity of the signatures. Both letters also signed by Young. MSS 624 A

1593. Seeler, Margaret. *Drawing of a mine.* 1965. 2 items.

One pencil drawing S. (1965), captioned "Old Colorado--[?--ealframe]; design for gallery's foot." The drawing contains a blank shield as if a coat of arms were to have been added, and holograph notes on how drawing should be reproduced. MSS 1350 A

1594. Bardeen, John, 1908- *Autograph signature.* 1966. 1 item (1 p.)

Autograph signature (26 Aug. [19]66) accompanying biographical typescript. MSS 46 A

1595. Enders, John Franklin, 1897- *Papers.* 1966. 2 items.

Autograph signature (1966 Aug. 9) with typed biography (1 p.), and A.L.S. (1966 Nov. 18) to Mr. Beau concerning Enders' inability to take time to answer various questions. MSS 483 A

1596. Giauque, William Francis, 1895- *Autograph signature.* 1966. 1 item (1 p.): col. ill.

Autograph signature on first day cover of Albert Einstein stamp (1966 Mar. 14, Princeton). MSS 592 A

1597. Javits, Jacob Koppel, 1904- *Letter.* 1966. 1 item (1 p.)

T.L.S. (1966 Sept. 1) to William J. Luetge concerning a commemorative stamp for Mr. Baruch. MSS 751 A

1598. Luria, Salvador Edward, 1912- *Autographed first day cover.* 1966. 1 item (1 leaf)

Autographed first day cover (1966 Mar. 14, Princeton) honoring Albert Einstein. MSS 931 A

1599. Richards, Dickinson W., 1895-1973. *Autograph signature.* 1966. 1 item (1 leaf)

Autograph signature (1966 Aug. 11) on page with T. description of his research leading to the Nobel Prize. MSS 1208 A

1600. Robinson, Robert, Sir, 1886-1975. *Letter.* 1966. 1 item (1 leaf)

T.L.S. (1966 July 18, London) to Dr. Laurent M. Lopez concerning his work on the chemistry of penicillin, specifically his editing the book *The chemistry of penicillin* and lecturing on "The early stages of the chemical study of penicillin," on letterhead of the Shell Centre. MSS 1224 A

1601. Rous, Peyton, 1879-1970. *Letter.* 1967. 1 item (1 leaf)

T.L.S. (1967 March 14, New York) to Marshall E. Bean regretfully refusing his request, on letterhead of The Rockefeller University. MSS 1239 A

1602. Duve, Christian de, 1917- *Autograph signature.* [between 1969 and 1977]. 1 item (1 p.)

Autograph signature with quotation on letterhead of The Rockefeller University.   MSS 416 A

1603. Goddard, Esther Christine Kisk. *Autograph signature.* 1969. 1 item (1 p.): col. ill.

Autograph signature on typed tribute of von Braun to Robert Goddard with a first day of issue stamp: "First man on the moon."   MSS 601 A

1604. Krebs, Hans Adolf, Sir, 1900- *Papers.* 1969. 2 items.

T.L.S. (1969 Nov. 26, Oxford) to Mrs. K. Janssen, enclosing a T. paper S. with extensive handwritten corrections later published in *Advances in Enzyme Regulation,* volume 8.   MSS 798 A

1605. Allen, Joseph P. *Autograph signature.* 1970. 1 item (1 p.)   MSS 73 A

1606. Wheatland, David P. *Electricity and magnetism before 1820 and other scientific books: short-title list of a collection formed by David P. Wheatland.* [1971?]. 252 [i.e., 253], [13] leaves; 29 cm.

Bound photocopied typescript followed by 13 leaves of photocopied typescript: "Additions to Books in the Collection--1968" and "Additions to the Book Collection as of January 1, 1971." Leaf 33 is lacking.   MSS 1307 B

1607. Van Allen, James Alfred, 1914- *Autograph signature.* [between 1972 and 1977]. 1 item (3 p.): ill.

Autograph signature on first day cover-brochure (1972 Feb. 14, United Nations, New York): Treaty on the non-proliferation of Nuclear Weapons.   MSS 1497 A

1608. Woodward, Robert Burns, 1917-1979. *Autograph signature.* 1973. 1 item (1 leaf)

Autograph signature on First Day Cover (1973 Apr. 23, Washington, D.C.) for 500th Anniversary of the Birth of Copernicus.   MSS 1586 A

1609. Beadle, George Wells, 1903- *Letter.* 1975. 1 item (1 p.)

A.L.S. (1975 April 19, Chicago) to Mr. Silverman regretting Beadle's inability to solve the problems of world hunger and overpopulation.   MSS 26 A

1610. Dibner, Bern, 1897- *Interviews.* 1977. 4 items.

Typed transcripts of 4 interviews given by Dibner to George Szabad in Norwalk, Connecticut.   MSS 438 A

1611. Dibner, Bern, 1897- *Interview.* 1979. 78 leaves; 29 cm.

Photocopy of typescript of an interview given by Bern Dibner to M. Krauss.   MSS 233 B

1612. Dibner Library. *Heralds of Science.* 1980. 18, [4], 9-92 leaves; ill.; 30 cm.

Typescript and photocopy corrected by M. Rosenfeld and E. Wells, spring 1980 for the revised edition.   MSS 253 B

[1613.] Brande, William Thomas, 1788-1866. *Summary.* [between 1800 and 1866]. 1 item (1 leaf)

Holograph leaf describing the divisions of a text (?) on electricity and the forces of nature, probably by William Thomas Brande. "Dr. Brande" written on verso.   MSS 1612 A

[1614.] *Letter to Dr. Charles Taylor.* 1806. 1 item (2 p.)

A.L. (posted 1806 Nov. 17, London?) written by an unidentified person to Dr. Charles Taylor, Secretary of the Society of Arts of Great Britain, concerning a meeting of the Society's Committee of Mechanics on F. C. Daniel's life preserver.   MSS 1613 A

Index